GOVERNING THE CLIMATE-ENERGY NEXUS

Combating climate change and transitioning to fossil-free energy are two central and interdependent challenges facing humanity today. Governing the nexus of these challenges is complex, and includes multiple intergovernmental and transnational institutions. This book analyzes the governance interactions between such institutions and explores their consequences for legitimacy and effectiveness. Using a novel analytical framework, the contributors examine three policy fields: renewable energy, fossil fuel subsidy reform, and carbon pricing. These fields are compared in terms of their institutional memberships, governance functions, and overarching norms. Bringing together prominent researchers from political science and international relations, the book offers an essential resource for future research and provides policy recommendations for effective and legitimate governance of the climate-energy nexus. Rooted in the most recent research, it is an invaluable reference for researchers, policy makers, and other stakeholders in climate change and energy politics.

FARIBORZ ZELLI is an associate professor in the Department of Political Science at Lund University. He is also a principal investigator and board member at BECC (Biodiversity and Ecosystem Services in a Changing Climate). His research focuses on international institutions, global environmental governance, and political theory and philosophy. He has edited four volumes: *Environmental Politics and Governance in the Anthropocene* (Routledge, 2016); *Encyclopedia of Global Environmental Governance and Politics* (Elgar, 2015); a special issue with *Global Environmental Politics* (2013); and *Global Climate Governance beyond 2012* (Cambridge University Press, 2010).

KARIN BÄCKSTRAND is a professor in environmental social science in the Department of Political Science at Stockholm University, where she codirects the Environmental Policy, Politics, and Learning (EPPLE) research group. Her primary research revolves around global environmental politics, the role of science in environmental decision making, the politics of climate change, and the democratic legitimacy of global governance. Her papers have been published in *Global Environmental Politics*, *European Journal of International Relations*, and *Global Environmental Change*. Her recent books are *The Research Handbook on Climate Governance* (Elgar, 2015) and *Rethinking the Green State* (Routledge, 2015).

NAGHMEH NASIRITOUSI is a postdoctoral research fellow at Stockholm University. Her research focuses on international climate change politics, energy governance, and issues of legitimacy and effectiveness. Her research has been published in journals such as *Nature Climate Change*, *European Journal of International Relations*, *International Studies Quarterly*, *Global Environmental Politics*, *International Environmental Agreements: Politics, Law, and Economics*, and *Annual Review of Political Science*.

JAKOB SKOVGAARD is an associate professor at Lund University and a principal investigator at BECC (Biodiversity and Ecosystem services in a Changing Climate). His research focuses on national, EU, and international climate politics – including the interactions between these levels and between economic and environmental objectives. He is coeditor of *The Politics of Fossil Fuel Subsidies and Their Reform* (Cambridge University Press, 2018) and of a special issue with *International Environmental Agreements: Politics, Law and Economics* (2017).

OSCAR WIDERBERG is an assistant professor at the Institute for Environmental Studies (IVM), Free University of Amsterdam. His teaching and research focus on international and transnational governance for climate change, biodiversity, and sustainability. He holds a PhD from the Free University of Amsterdam. Before joining academia, he worked in international public policy consulting.

"Far too little work looks at climate change or energy studies from an interdisciplinary social science perspective, and far less than that does so effectively. This book is a notable exception, expertly tracing the governance dynamics and dilemmas of climate change as well as its twin partner in crime, the global energy system. In doing so, it covers a range of exciting topics, from subsidy reform to institutional complexity, from renewable energy to policy coherence, through a compelling milieu of different perspectives. The book is truly important reading for anyone who professes to take the study of climate governance seriously."

– Benjamin K. Sovacool, University of Sussex, and author of Global Energy Justice *(Cambridge University Press)*

"If we are to prevent catastrophic global warming, the world needs to establish more effective global governance of energy. This carefully planned book brings together contributions by leading experts and provides an insightful macro perspective on the current climate-energy nexus – its institutional complexity and fragmentation, as well as the potential for change. The volume stands out for its rich empirical analysis, coherence and rigour. A must read!"

– Robert Falkner, London School of Economics and Political Science

"Governing the Climate-Energy Nexus offers a novel approach to understanding the vexing challenge of decarbonizing the global energy system and moving toward a more sustainable future. Drawing on diverse theoretical debates in political science and international relations, the authors present rich empirical analyses that help academics and practitioners navigate the complex institutional landscape of global climate and energy governance and evaluate the trade-offs and synergies between different policy options. Moreover, this book advances the growing field of 'nexus' research by breaking new theoretical and methodological ground that will facilitate more effective and legitimate governance systems in an increasingly interconnected global system."

– Michele M. Betsill, Colorado State University

GOVERNING THE CLIMATE-ENERGY NEXUS

Institutional Complexity and Its Challenges to Effectiveness and Legitimacy

Edited by

FARIBORZ ZELLI
Lund University

KARIN BÄCKSTRAND
Stockholm University

NAGHMEH NASIRITOUSI
Stockholm University

JAKOB SKOVGAARD
Lund University

OSCAR WIDERBERG
Free University of Amsterdam

CAMBRIDGE
UNIVERSITY PRESS

CAMBRIDGE
UNIVERSITY PRESS

University Printing House, Cambridge CB2 8BS, United Kingdom

One Liberty Plaza, 20th Floor, New York, NY 10006, USA

477 Williamstown Road, Port Melbourne, VIC 3207, Australia

314–321, 3rd Floor, Plot 3, Splendor Forum, Jasola District Centre, New Delhi – 110025, India

79 Anson Road, #06–04/06, Singapore 079906

Cambridge University Press is part of the University of Cambridge.

It furthers the University's mission by disseminating knowledge in the pursuit of education, learning, and research at the highest international levels of excellence.

www.cambridge.org
Information on this title: www.cambridge.org/9781108484817
DOI: 10.1017/9781108676397

First published 2020

A catalogue record for this publication is available from the British Library.

Library of Congress Cataloging-in-Publication Data
Names: Zelli, Fariborz, editor. | Bäckstrand, Karin, editor. | Nasiritousi, Naghmeh, editor. | Skovgaard, Jakob, 1977– editor. | Widerberg, Oscar, editor.
Title: Governing the climate-energy nexus : institutional complexity and its challenges to effectiveness and legitimacy / Edited by Fariborz Zelli, Lund University, Karin Bäckstrand, Stockholm University, Naghmeh Nasiritousi, Stockholm University, Jakob Skovgaard, Lund University, Oscar Widerberg, VU University Amsterdam.
Description: Cambridge ; New York : Cambridge University Press, 2020. | Includes bibliographical references and index.
Identifiers: LCCN 2019039822 (print) | LCCN 2019039823 (ebook) | ISBN 9781108484817 (hardback) | ISBN 9781108676397 (epub)
Subjects: LCSH: Global environmental change–Government policy. | Fossil fuels–Government policy. | Renewable energy sources–Government policy.
Classification: LCC GE149 .G67 2020 (print) | LCC GE149 (ebook) | DDC 393.738/74561–dc23
LC record available at https://lccn.loc.gov/2019039822
LC ebook record available at https://lccn.loc.gov/2019039823

ISBN 978-1-108-48481-7 Hardback

Contents

Contributors

Karin Bäckstrand is Professor of Environmental Social Science at the Department of Political Science, Stockholm University, Sweden.

Jana Canavan is a PhD candidate at the Department of Political Science, Lund University, Sweden.

Klaus Hammes, PhD, is Chief Economist at the Swedish Energy Agency.

Naghmeh Nasiritousi, PhD, is Research Fellow at the Department of Political Science, Stockholm University, Sweden.

Philipp Pattberg is Professor of Transnational Environmental Governance and Policy, Vrije Universiteit Amsterdam, The Netherlands.

Lisa Sanderink is a renewable energy advisor at the Netherlands Enterprise Agency and a PhD candidate at the Institute for Environmental Studies, Vrije Universiteit Amsterdam, The Netherlands.

Jakob Skovgaard is an associate professor at the Department of Political Science, Lund University, Lund, Sweden.

Harro van Asselt is Professor of Climate Law and Policy at the University of Eastern Finland Law School, Finland.

Soetkin Verhaegen, PhD, is a research fellow at the Department of Political Science, Stockholm University, Sweden, and affiliated researcher at the Institute of Political Science Louvain-Europe (ISPOLE), Université Catholique de Louvain, Belgium.

Cleo Verkuijl is a research fellow at the Stockholm Environment Institute, Oxford, and an adjunct professor at Johns Hopkins University, Bologna.

Oscar Widerberg is an assistant professor at the Institute for Environmental Studies, Vrije Universiteit Amsterdam, The Netherlands.

Fariborz Zelli is an associate professor at the Department of Political Science, Lund University, Sweden.

Foreword

This book is a major outcome of an international research project – *Challenges and Opportunities in a Fragmented Global Energy and Climate Governance* 2015–2018 (CLIMENGO) – funded by the Swedish Energy Agency. In my view, this volume gives an excellent overview of the climate-energy nexus in a global perspective. I have learned a lot that is useful for my work in the Swedish Energy Agency, and the same can be said most likely for civil servants, experts, and policy makers in other public agencies. This book may also help those that consider starting a new organization to identify important gaps to fill, and to navigate in the context of institutional complexity that characterizes the fields of climate and energy governance.

Why study the climate-energy nexus? The Paris Agreement is the latest major intergovernmental treaty that calls for urgent action on climate change, while we know that energy stands for around two-thirds of greenhouse emissions worldwide. It is hence vital to understand which international organizations work in the intersecting fields of climate and energy governance, how the multitude of international institutions is structured and perceived by various stakeholders, and where crucial gaps exist with regard to governance, effectiveness, and legitimacy. The book provides a very valuable and novel set of empirical research findings on these questions: about the plethora of actors and institutional complexity in the climate-energy nexus, about what institutions and actors do, about what they don't do, and how this matters for the performance and perception of climate and energy governance.

The volume starts with a clear theoretical framework and methodology on how to analyze the climate-energy nexus through four interacting variables – coherence, management, effectiveness, and legitimacy. In the following chapters this framework is applied to the macro, meso, and micro levels for three selected case studies: renewable energy, carbon pricing, and fossil fuel subsidy reform. The contributions address major research gaps and identify important connections between the plurality of institutions and actors within and across these governance subfields. Thereby, these case studies provide an integrated mapping and a consistent analysis of the institutional complexes governing the nexus between climate change and energy.

The understanding of the nexus is further enhanced by other features of the case studies, for example, the timelines and accounts of institutional development that are provided for renewable energy, fossil fuel subsidies reform, and carbon pricing. It is very insightful to see how early some policy areas emerged and began to be populated by institutions. This goes in particular for the subfield of renewable energy, with the International Solar Energy Society as the very first institution, which began operating several decades prior to the commercial availability and affordability of photovoltaic cells. On the other hand, we learn from the case studies about the limited support that other areas have received so far, especially a critical theme such as the reduction or abolishment of fossil fuel subsidies – and this, although international organizations such as the International Energy Agency had repeatedly pointed out the importance of this issue for achieving a reduction of carbon emissions in line with the goals of the Paris Agreement.

In addition to these important and highly insightful mappings and analyses, the volume includes impressive efforts to examine the performance of the climate-energy nexus. To this end, two chapters evaluate the effectiveness and legitimacy of various institutions in this nexus under conditions of institutional complexity.

In light of these achievements and novel findings, I really enjoyed reading this book and I hope many will share my experience.

Klaus Hammes, Chief Economist, Swedish Energy Agency

Preface and Acknowledgements

Global climate and energy governance increasingly intersect. Transformation in the consumption and production of energy across developing and developed countries is key to setting the world on a low-carbon path. It therefore makes sense, as we argue in this book, to talk about a global climate-energy nexus. This nexus consists of a plethora of international institutions that govern numerous policy fields, such as renewable energy, fossil fuel subsidies, and carbon pricing. These issues, and the institutions that govern them, all have far-reaching consequences for climate change and energy use, and thereby for the future world we live in.

Many of these institutions overlap in their mandates and memberships of governments and nongovernmental actors, while they exercise different governance functions. All this leads to a high degree of institutional complexity across the climate-energy nexus. We lack a deeper understanding of the shapes of this complexity, and of its crucial consequences for the legitimacy and effectiveness of the governance arrangements that are operating therein.

This book provides essential answers to these research gaps. It conceptualizes the governance of the climate-energy nexus and presents novel empirical findings through carefully crafted comparative case studies that combine different methodologies in an innovative way.

The volume is the outcome of a four-year international research project on *Challenges and Opportunities in a Fragmented Global Energy and Climate Governance* 2015–2018 (CLIMENGO), funded by the Swedish Energy Agency (SEA). Coordinated by Karin Bäckstrand at Stockholm University, with collaborative partners at Lund University and Vrije University Amsterdam, it has united leading experts in the fields of energy and climate governance. We are immensely grateful to the SEA for giving us the opportunity to carry out the important and innovative research that we present in this book. We are also thankful for the possibility to present our research at the annual SEA conferences, and we are particularly indebted to Marie Carlberg, Klaus Hammes, Josefin Thoresson, and Maria Vuorelma.

The idea to produce a joint edited volume arose in early 2018. In June 2018, we could already discuss first draft chapters at a two-day workshop, which was kindly hosted by Harro van Asselt and Cleo Verkuijl at the Oxford branch of the Stockholm Environment Institute. The process of writing this book over the following year has been a stimulating, fruitful, and intense academic journey and conversation, which has cut across and bridged different scholarly fields such as international relations, environmental politics, law, geography, and environmental sciences. We presented our revised book chapters in a panel organized at the annual meeting of International Studies Association (ISA) in Toronto in March 2019 and want to express our gratitude to our discussant Matt Hoffmann from the University of Toronto. He provided extremely useful and excellent feedback that helped us to enhance the contributions to this book during the final steps before the submission of our manuscript.

As editors, we have been privileged to work with an exceptional group of authors who have contributed to this volume: Jana Canavan, Phillip Pattberg, Lisa Sanderink, Harro van Asselt, Cleo Verkuijl, and Soetkin Verhaegen. We extend our sincere gratitude to them for their collegial spirit, critical input, helpful advice, and tremendous patience – not only with respect to their own chapters, but for all of their invaluable support in improving the quality of the entire book.

Likewise, this volume would not have been possible without the helpful input from a wide range of people. This begins with the editorial team at Cambridge University Press, in particular Lisa Bonvissuto, Sarah Lambert, and Matt Lloyd, who have supported us with their great dedication and guidance throughout all the production phases of this volume. We also want to take this opportunity to thank the three reviewers for Cambridge University Press. They challenged us to further clarify and sharpen the book's theoretical, methodological, and empirical added value, helping us to advance the research frontier on the climate-energy nexus.

Moreover, we would like to express our thanks to a number of colleagues, who have dedicated their time to provide valuable feedback at various stages over the past four years. They all helped us to make significant progress in our research project and to accomplish this volume. This includes Jonas Tallberg and Lisa Dellmuth of the University of Stockholm, who offered constructive comments on the legitimacy aspects of institutional complexity. A special thanks goes to two persons who were involved in the early stages of the project as well as the application phase: Daniel Stenson Engström at the Swedish Environmental Protection Agency, and Jonathan Kuyper at Queens University Belfast. Other colleagues have contributed in various ways to specific chapters, notably Jasmiini Pylkkänen from the University of Oulu.

Finally, we are indebted to our research assistant Hugo Faber at Stockholm University, whose dedicated and meticulous editorial assistance as well as interview transcriptions were invaluable.

As editors, we take full responsibility for any errors or omissions remaining in the book.

Abbreviations

ACA	Airport Carbon Accreditation
ACE	Association of Southeast Asian Nations Centre for Energy
AEEP	Africa–EU Energy Partnership
APEC	Asia Pacific Economic Cooperation
APEC_EWG	Asia Pacific Economic Cooperation Energy Working Group
AREI	Africa Renewable Energy Initiative
ASEAN	Association of Southeast Asian Nations
BASREC	Baltic Sea Region Energy Cooperation
BTEC	Break Through Energy Coalition
C4C	United Nations Global Compact Caring for Climate
CAA	Clean Air Asia
CATF	Climate Action Takes Flight
CCREEE	Caribbean Center for Renewable Energy and Energy Efficiency
CDM	Clean Development Mechanism
CEM	Clean Energy Ministerial
CESC	Clean Energy Solutions Center
CIF	Climate Investment Funds
CLASP	The Collaborative Labelling and Appliance Standards Program
CLIMATE_ALLIANCE	Climate Alliance of European Cities with Indigenous Rainforest Peoples
CNCA	Carbon Neutral Cities Alliance
CNP	Carbon Neutral Protocol
COM	Covenant of Mayors

CORSIA	Carbon Offsetting and Reduction Scheme for International Aviation
CPLC	Carbon Pricing Leadership Coalition
CSR	Corporate Social Responsibility
CTIPFAN	Climate Technology Initiative of the Private Financing Advisory Network
CUD	Connected Urban Development
DI	Divest Invest Global Movement
E4I	Energy for Impact (formerly GVEP)
ECO	ECO Partnerships Clean Energy and Energy Efficiency
ECREEE	Economic Community of West African States Center for Renewable Energy and Energy Efficiency
EN_CITIES	Energy Cities
ENERGIA	International Network on Gender and Sustainable Energy
ENERGY+	International Energy and Climate Initiative Energy Plus
ENLIGHT	En.Lighten Initiative
ENR	European Energy Network
ENTSOE	European Network of Transmission System Operators
ETS	Emission Trading System
EU	European Union
EUEI	European Union Energy Initiative
EUROCITIES	EUROCITIES
EUROSOLAR	The European Association for Renewable Energy
EWG	Energy Working Group
FFC	Fleets for Change
FFSR	Fossil Fuel Subsidy Reform
Friends	Friends of Fossil Fuel Subsidy Reform
G7	Group of Seven
G20	Group of Twenty
G20_SR	Group of Twenty Subsidy Reform
GACC	Global Alliance for Clean Cookstove
GBEP	Global Bioenergy Partnership
GEEAP	Global Energy Efficiency Accelerator Platform
GEEREF	Global Energy Efficiency and Renewable Energy Fund
GFA	Green Freight Asia
GFAAF	Global Framework for Aviation Alternative Fuels
GFE	Green Freight Europe

GFEI	Global Fuel Economy Initiative
GFF	Go Fossil Free
GGFAP	Global Green Freight Action Plan
GGFRP	Global Gas Flaring Reduction Partnership
GMI	Global Methane Initiative
GNESD	Global Network on Energy for Sustainable Development
GS	The Gold Standard
GSA	Global Solar Alliance
GSC	Global Solar Council
GSEP	Global Sustainability Electricity Partnership (formerly the E8)
GSI	Global Subsidies Initiative
HLPF	High-Level Political Forum on Sustainable Development
IATA	International Air Transport Association
IATA_COP	International Air Transport Association Carbon Offset Program
ICAO	International Civil Aviation Organization
ICAP	International Carbon Action Partnership
ICLEI	Local Governments for Sustainability
IEA	International Energy Agency
IEEA	Industrial Energy Efficiency Accelerator
IETA	International Emissions Trading Association
IMF	International Monetary Fund
INFORSE	International Network for Sustainable Energy
IPEEC	International Partnership for Energy Efficiency Cooperation
IRENA	International Renewable Energy Agency
ISCI	International Solar Cities Initiative
ISES	International Solar Energy Society
ISGAN	Implementing Agreement for a Cooperative Program on Smart Grids
JI	Joint Implementation
KP	Kyoto Protocol
L&G	Lean and Green
LCTPI	Low Carbon Technology Partnerships Initiative
LEDS_GP	Low Emission Development Strategies Global Partnership
MEF	Major Economies Forum
MI	Mission Innovation
NCM	Networked Carbon Markets Initiative

NDC	Nationally Determined Contribution
NEG_ECP	New England Governors and Eastern Canadian Premiers' Annual Conference
NGO	Nongovernmental Organization
OECD	Organisation for Economic Co-operation and Development
OLADE	Latin American Energy Organization
OPEC	Organization of the Petroleum Exporting Countries
PCFV	United Nations Environment Programme Partnership for Clean Fuels and Vehicles
PDC	Portfolio Decarbonization Coalition
PMR	Partnership for Market Readiness
PPMC	Paris Process on Mobility and Climate
R20	Regions of Climate Action
RE100	100 Percent Renewables
REC	Renovate Europe Campaign
RECP	Africa–EU Renewable Energy Cooperation Program
REEEP	Renewable Energy and Energy Efficiency Partnership
REN21	The Renewable Energy Policy Network for the Twenty-First Century
RN!	Refrigerants, Naturally!
RSB	The Roundtable on Sustainable Biofuels (RSB Standard)
SBCI	United Nations Environment Programme Sustainable Buildings and Climate Initiative
SDG	Sustainable Development Goal
SE	Shipping Efficiency
SEAD	Super-Efficient Equipment and Appliance Deployment Initiative
SEADS	European Union Energy Initiative Strategic Energy Advisory and Dialogue Service
SEforALL	Sustainable Energy for All
SIDS_DOCK	Small Island Developing States Sustainable Energy and Climate Resilience Initiative
SLOCAT	Partnership on Sustainable Low Carbon Transport
TCC	Think Climate Coalition
U4E	United for Efficiency
UEMI	Urban Electric Mobility Initiative
UN	United Nations
UN_EN	United Nations Energy

UNEP	United Nations Environment Programme
UNFCCC	United Nations Framework Convention on Climate Change
VCS	Verified Carbon Standard (formerly the Voluntary Carbon Standard)
VER	Verified Emissions Reductions
VER+	VER Plus
VPR	Voluntary Peer Review
WBCSD_E&C	World Business Council for Sustainable Development Energy and Climate
WCI	Western Climate Initiative
WGBC	World Green Building Council
WTO	World Trade Organization
WWF	World Wildlife Fund
ZEV	International Zero-Emission Vehicle Alliance

1

Introduction

Governing the Climate-Energy Nexus

FARIBORZ ZELLI, KARIN BÄCKSTRAND, NAGHMEH NASIRITOUSI, JAKOB SKOVGAARD, AND OSCAR WIDERBERG

Energy and climate change are fundamentally connected. In today's world energy production and use account for two-thirds of global greenhouse gas emissions (IEA 2018). To curb the climate crisis and to meet the goals set out in the Paris Agreement, we need to provide reliable and affordable energy to some 10 billion people. This much needed transition to low- or zero-carbon societies would have profound consequences for the world's energy systems (Lesage and Van de Graaf 2016).

Many steps have already been taken to bring this transition on its way, and they include an ever-increasing number of governance initiatives across borders. The success of this transition – and therewith the future quality of life on our planet – depends on how effectively and legitimately these various governance efforts achieve their goals, and this without undermining each other too much. In short, coordinated policy responses to limit climate change and decarbonize energy systems worldwide need to go hand in hand.

Against the backdrop of this urgent link between global climate change governance and global energy governance, this volume puts the 'climate-energy nexus' at the forefront. Both governance structures making up this nexus are highly complex (Sovacool 2013; Goldthau et al. 2018). Efforts to tackle climate change have increased significantly over the past twenty and more years, and they have expanded far beyond the multilateral response under the United Nations Framework Convention on Climate Change (UNFCCC) (Vogler 2018). These efforts vary from minilateral, local, and transnational initiatives to private regulation, technology agreements, and market-based mechanisms (Hoffman 2011; Hjerpe and Nasiritousi 2015). Likewise, governance efforts to address energy demands on the international level are fragmented, and even lack a core multilateral institution similar to the role the UNFCCC plays within the institutional complex on climate change (Sovacool and Florini 2012; Wilson 2015; Van de Graaf and Colgan 2016; Sovacool and Van de Graaf 2018).

This fragmented climate-energy nexus may imply severe challenges for both state and nonstate actors to effectively combat climate change and transition to fossil-free energy systems (Falkner 2014). There is limited knowledge to which degree the fragmentation translates into (in)coherence, and what kind of consequences the degree of coherence has on the effectiveness and legitimacy of governance efforts. Apart from the number of institutions a range of other dimensions are also at play, including the kinds of actors involved, interpretations of core norms and distribution of governance functions (see Chapter 2; Biermann et al. 2009; Eberlein et al. 2014). One important aspect that has been sidelined in the literature is that governance *in* the climate-energy nexus is rarely governance *of* the climate-energy nexus in its entirety: Governance in the nexus mainly targets subfields focusing on specific issues such as renewable energy, carbon pricing, or fossil fuel subsidies.

This book aims to address this research gap and identify connections between the plethora of institutions and actors that address climate change and energy challenges within and across specific governance subfields. First, it provides novel mappings of the institutional complex of the climate-energy nexus as a whole and of the respective complexes governing three major policy subfields within the nexus – renewable energy, fossil fuel subsidy reform, and carbon pricing. Second, and based on this mapping, the volume analyzes the degrees of institutional coherence for all three subfields and studies to what extent there have been deliberate management attempts targeting the interlinkages between institutions. Third and finally, the consequences of the multifaceted institutional complex in the climate-energy nexus are examined by asking how the institutional complexity affects perceptions of effectiveness and legitimacy across the selected subfields.

The next three sections explain the rationale and theoretical and empirical contributions of the edited volume and our motivation to select three specific policy subfields as case studies. The introduction concludes with an outline of the chapters, structured along the book's three parts on mapping, coherence and management, and legitimacy and effectiveness.

1.1 Rationale behind This Book

A defining feature of global environmental governance today is the patchwork of overlapping institutions with varying forms and functions that govern different aspects of environmental challenges (cf. Young 1996; Oberthür 2009; Zelli and van Asselt 2013; van Asselt and Zelli 2014; Heubaum and Biermann 2015; Pattberg and Widerberg 2015; Boas et al. 2016). A case in point is the intersection of global climate and energy governance.

International, transnational and nonstate action to mitigate climate change has proliferated in number and scope in the lead-up to the Paris Agreement (Bulkeley et al. 2014; Jordan et al. 2015; Bäckstrand et al. 2017; van Asselt and Zelli 2018). For instance, the Non-State Actor Zone for Climate Action (NAZCA) online portal,[1] maintained by the United Nations Framework Convention on Climate Change (UNFCCC) Secretariat, has identified more than seventy cooperative initiatives involving governments (of which more than twenty specifically target energy), in addition to thousands of nonstate and subnational climate actions.

At the same time, national sovereignty concerns have left multilateral cooperation on energy without a central or focal institution similar to the UNFCCC in climate governance (Karlsson-Vinkhuyzen 2010; Sovacool and Florini 2012; Barnsley and Ahn 2014; Wilson 2015; Van de Graaf and Colgan 2016; Sovacool and Van de Graaf 2018). As current energy systems are a major source of greenhouse gas emissions, transformation in the consumption and production of energy in major emitter states, across developing and developed countries, is a key in setting the world on a low-carbon path (Goldthau and Witte 2010; Cherp et al. 2011; Dubash and Florini 2011; Nasiritousi 2017; IEA 2018). Hence, the socioeconomic and biophysical interactions between climate and energy are very strong.

In this volume, three central issues at the intersection of climate and energy governance have been selected: renewable energy, fossil fuel subsidy reform, and carbon pricing. Comparing these three policy areas, it can be observed that they are characterized by multiple governance institutions and a patchwork of actors, rules, and decision-making processes (Van de Graaf and Colgan 2016; Sanderink et al. 2017). Renewable energy, to take one example, is discussed not only within the United Nations (UN), but also in technology partnerships and international organizations such as the International Renewable Energy Agency (IRENA), the Clean Energy Ministerial (CEM), the Climate Technology Initiative (CTI), Energy +, and in a number of private–public partnerships and private initiatives such as the Renewable Energy Policy Network for the 21st Century (REN21) (Widerberg and Pattberg 2015; Sanderink et al. 2017; Sovacool and Van de Graaf 2018). Many of these institutions have overlapping mandates and memberships among governments and nongovernmental actors, while exercising different governance functions (Keohane and Victor 2011; see also Biermann et al. 2009).

Institutional complexity in general as well as the relationship between climate and energy governance in particular are attracting growing interest among scholars (e.g. Bäckstrand 2008; Zelli 2011; Van de Graaf 2013; Widerberg and

[1] https:climateaction.unfccc.int/ (accessed 23 February 2018).

Stenson 2013; van Asselt 2014; Falkner 2018; Goldthau et al. 2018). Their research has shown that the increasing number of partnerships in climate and energy governance has contributed to the fragmentation of the institutional landscape (Biermann 2009; Van de Graaf 2013; Zelli et al. 2013; Jordan et al. 2015). Terms such as 'regime complex' (Keohane and Victor 2011), 'polyarchy' (Ostrom 2010), 'fragmented governance architectures' (Biermann et al. 2009), and 'climate anarchy' (Dyer 2014) equally reflect this development in current global environmental governance. And significant research efforts have demonstrated that this complexity also marks the intersections between policy domains, for instance, between the global governance of climate change and other policy areas such as trade (e.g. Charnovitz 2003; van Asselt 2014), biodiversity (e.g. Jacquemont and Caparrós 2002), or development and security (e.g. Moncel and van Asselt 2012).

Beyond the institutional complexity literature, the book also contributes to discussions around various 'nexuses' in public policy objectives. Over the past fifteen years, the nexus concept has become a popular way to illustrate connectivity between various policy fields. The 'water-food-energy' nexus has been a particular focus of attention (e.g. Hoff 2011) but other combinations abound. The nexus concept has been used to connect just about any policy field, including, to name but a few, climate change, gender, health, poverty, education, and agriculture (e.g. Clancy et al. 2002; Iguchi et al. 2014). Thinking in terms of nexuses among policy fields – instead of individual policy fields – is expected to generate insight into trade-offs and synergies between various policy decisions, or, as the UN's Food and Agricultural Organization (2014, 4) puts it: a nexus approach '*forces us to think of the impacts a decision in one sector can have not only on that sector, but on others. Anticipating potential trade-offs and synergies, we can then design, appraise and prioritise response options that are viable across different sectors*'. Policies to address climate change using forests as carbon sinks, for instance, may have impacts on biodiversity, agriculture, and water and they require decision makers to take an integrated approach toward sustainability challenges. For researchers studying a nexus, enhancing policy coherence is hence key to reap the benefits of potential synergies and mitigate negative effects across various policy goals (Munaretto and Witmer 2017).

The achievements of these different literatures on climate change, energy, complexity, and nexuses notwithstanding, there are several research questions that have not been comprehensively addressed so far. First, there are few attempts to analyze the nexus between climate change and energy by comparing the institutional complexes for specific subfields (e.g. Bazilian et al. 2011; Sovacool 2011; Bradshaw 2014; Falkner 2014, 2018; Van de Graaf and Colgan 2016). Second, while all studies share the starting assumption that institutional complexity matters, there are no systematic comparative analyses about how exactly it may affect the effectiveness or legitimacy of individual institutions or entire institutional complexes, or how certain actors may benefit or be disadvantaged in a complex governance architecture (Andresen and Hey 2005; Karlsson-Vinkhuyzen and McGee 2013; Bäckstrand et al. 2018; Sovacool and Van de Graaf 2018). Third, in empirical terms, the scholarship on institutional interactions has mostly provided insights into dyadic relationships between two international regimes or public institutions composed by national governments (Oberthür and Gehring 2006; Chambers 2008; Young et al. 2008; Gehring and Oberthür 2009; Oberthür and Stokke 2011), while placing less emphasis on broader governance interactions that also include hybrid, private, and informal institutions (Heubaum and Biermann 2015; Sanderink et al. 2017).

In summary, there is a gap in knowledge of the nature of the relationship between institutions governing the climate-energy nexus in a multilevel context as well as the implications of that institutional complexity. This book seeks to address these gaps.

1.2 Objectives and Contribution: Analyzing the Climate-Energy Nexus

The aim of this volume is to comprehensively map, critically analyze, and compare a wide range of interactions among intergovernmental and transnational institutions across three policy fields in the climate-energy nexus.

Conceptually, this book advances the research frontier in the scholarship on institutional complexity and fragmentation (Zelli and van Asselt 2013) by zooming in on governance interaction and policy integration between state and nonstate actors. The linkages between institutions are conceptualized along four evaluative themes and their respective dimensions: (1) coherence and (2) management of institutional complexes as well as their consequences for the (3) legitimacy and (4) effectiveness of institutions within institutional complexes.

Empirically, the contributions to this book break important new ground, as they provide novel findings on the institutional complexes that govern three central policy domains: renewable energy, fossil fuel subsidy reform, and carbon pricing.

Theoretically, the book explores how the position of a policy field in a nexus may influence the degree of complexity between institutions.

The following research questions guide this book:

Mapping: What is the institutional structure addressing the climate-energy nexus? Which types of institutions and actors are involved, which governance functions do they perform, and which specific policy field (e.g. renewable energy) do they address?

Coherence and Management: Are the relations between institutions in different subsets of the climate-energy nexus conflictive or synergistic? To what extent are there convergence, divergence or governance gaps with regard to overarching norms, governance functions, and membership? Are there deliberate attempts to address shortcomings of cross-institutional relations by certain actors?

Legitimacy and Effectiveness: What do the interactions between various institutions in global climate and energy governance mean for the legitimacy, legitimation, and effectiveness of individual institutions and the climate-energy nexus as whole?

We address these questions in three consecutive parts: In Part I, we introduce the conceptual and analytical framework and provide comprehensive mappings of the institutional structure of the climate-energy nexus as a whole. Part II provides in-depth analyses on the coherence and management of three policy domains of the climate-energy nexus: those addressing renewable energy, the phasing out of fossil fuel subsidies, and carbon pricing. Part III presents comparative studies that examine whether, and in which ways, different types of institutions working in the same policy field are perceived as legitimate or effective.

Across these three parts, the analyses build on a mixed-methods approach, including content analysis, network analysis, surveys, and semi-structured interviews with key stakeholders and experts. Moreover, all three parts provide a multi-level governance perspective on the climate-energy nexus. While the book departs from an analysis of the global governance of the nexus, several chapters examine roles and perceptions at the domestic and sub-state levels. On the one hand, preferences of member states (such as China, USA, EU) and energy industry actors set the agenda for global institutions. On the other hand, governments and stakeholders are affected by the high level of institutional complexity of global climate and energy governance. Some actors are better equipped than others to navigate the complex institutional system and, likewise, their perceptions of the legitimacy and effectiveness of certain institutions may differ considerably.

1.3 Three Case Studies: Renewable Energy, Fossil Fuel Subsidy Reform, and Carbon Pricing

We selected three policy subfields within the climate-energy nexus based on three main criteria. The first of these is the urgency and importance of each subfield in decarbonizing energy systems and combating climate change.

The subfield of renewable energy is key for a sustainable future, as the global uptake of renewable energy plays a central role in the decarbonization of global energy systems (Röhrkasten 2018). As Chapter 4 by Sanderink lays out in further detail, renewable energy is crucial for addressing three closely connected energy goals simultaneously: to provide energy security by satisfying increasing energy demand; to ensure worldwide energy access, with currently 1.1 billion people not having access to electricity; and to tackle negative environmental externalities associated with today's energy systems, with climate change being arguably the most important of these.

Fossil fuel subsidy reform (FFSR) covers the complex of institutions that seek to rationalize and phase out inefficient subsidies in the medium term (Chapter 5 by Verkuijl and van Asselt). Such subsidies potentially lock in an unsustainable fossil fuel infrastructure for years to come. According to estimates, more than a third of carbon emissions between 1980 and 2010 were driven by fossil fuel subsidies (Stefanski 2016), and phasing them out could at least reduce carbon emissions by 10 per cent (UNEP 2018).

The rationale of the subfield of carbon pricing is that climate change is best mitigated by giving emitters an incentive to reduce emissions through price signals, with the decision of how to reduce emissions best left to the market (Chapter 6 by Skovgaard and Canavan). With a few exceptions, emissions covered by carbon pricing arise from energy use (within industry, transportation and power generation), and carbon prices therefore help direct the choice of fuels away from emissions-intensive ones.

Second, as Chapters 4–6 demonstrate, these three policy subfields vary considerably in the number and mix of institutions that govern them at the international level. With this cross-case variation as a second selection criterion, the book provides an insightful comparison of the very different institutional complexes, their coherence and management, and the various consequences for legitimacy and effectiveness.

The renewable energy subfield is governed by a wide range of different institutions, including intergovernmental organizations alongside private institutions and multi-stakeholder partnerships. This notwithstanding, the subfield is still dominated by national policy-making as nation states continue to have sovereign control over the energy domain. By contrast, the carbon pricing subfield is predominantly governed on the international level and the respective institutional complex exhibits a medium number of public, private, and hybrid institutions. Membership in the few existing international FFSR institutions, finally, is heavily skewed toward public institutions, while the role of private and hybrid institutions is limited for this subfield.

A third and final case-selection criterion relates to the variation within the thematic structure in which a problem or policy field is embedded. In this volume, the climate-energy nexus constitutes this structure. In other words, we take into account the position of a policy subfield within the climate-energy nexus. Positioning is here understood as the degree to which the theme of a subfield is defined as a climate or energy problem, as well as the extent to which policy measures within the field (intentionally or not) influence carbon emissions and energy use.

The three selected policy fields indeed differ in their positioning within the climate-energy nexus. Carbon pricing is mainly related to climate change, since the notion of pricing carbon emissions is based on the definition of climate change as a policy problem. This said, while carbon pricing only aims to address climate mitigation, it mainly targets emissions from energy use (Métivier et al. 2018; Postic and Métivier 2019), and induces less energy consumption and a shift from high emission to lower emission energy sources. Fossil fuel subsidy reform is arguably as much a climate change policy problem as an energy issue. The subsidies as such target energy use, but fossil fuel subsidy reform is increasingly defined as a climate mitigation instrument (Terton et al. 2015; Jewell et al. 2018; Van de Graaf and Blondeel 2018). Finally, renewable energy is rooted in the energy domain, alhough one of the main purposes of promoting renewable energy is reducing carbon emissions.

The positioning of the policy fields within the climate-energy nexus matters, inter alia, due to the different governance architectures addressing climate change and energy. Whereas climate change is governed by a fragmented system with one central institution, the UNFCCC (Biermann et al. 2009), energy is governed by a fragmented and decentralized governance architecture without one central institution (Van de Graaf and Colgan 2016). Consequently, renewable energy operates further from the gravitational influence of the UNFCCC than fossil fuel subsidy reform and especially carbon pricing.

In the conclusions, this problem-structural argument and selection criterion will be revisited. We will discuss to what extent the different nature of the three problems is mirrored in different institutional architectures and different consequences for legitimacy and effectiveness.

Having introduced our rationale behind selecting the three policy subfields, we like to make one important qualification: the delineation of a governance problem or policy field has an unavoidable element of construction to it, since it depends on how the observers define the overarching norm, goal, or essence of the field in question. In formulating the core norms for each of the three subfields (see Chapters 4–6 for elaborate definitions of the respective norms), the authors therefore made necessary subjective assessments of what counts as a subfield within the climate-energy nexus.

1.4 Structure of the Book

1.4.1 Mapping the Climate-Energy Nexus

Part I consists of two chapters that provide analytical guidance and overarching empirical mappings. Chapter 2, by Zelli, Nasiritousi, Bäckstrand, Pattberg, Sanderink, Skovgaard, van Asselt, and Widerberg, introduces the overarching analytical framework for this book. The chapter first establishes our understanding of key terms such as institutions, institutional complex, nexus, and global governance. In a second step, it distinguishes three analytical levels for investigating institutional complexity: the macro level, which captures the overarching institutional fragmentation for a given field or domain of global governance; the meso level, which zooms into specific subfields of such a domain and the institutional complexes thereof; and the micro level, which exhibits interlinkages among two individual institutions on specific questions.

Chapter 2 then establishes the four evaluative themes that are employed in Parts II and III: coherence, management, legitimacy, and effectiveness. For each of these themes a series of dimensions is introduced to guide the examination of the meso and micro levels of a policy field. Coherence among institutions, for instance, is conceptualized in terms of: adherence to and interpretation of an overarching core norm for the policy field, coverage and distribution of memberships (private, public, hybrid), coverage and distribution of governance functions (standards and commitments, operational activities, information and networking, financing), and mechanisms underlying cross-institutional relations (cognitive, normative, behavioural). Legitimacy, to give another example, is to be assessed along nine dimensions, among them expertise, transparency, accountability, procedural fairness and distributive fairness. Altogether, the four themes and their dimensions make up a novel framework for an in-depth analysis of a governance nexus, such as the one on energy and climate change. They help us examine a variety of important questions in a comparative research design, combining a high level of ambition with feasibility and novelty.

Chapter 3, by Sanderink, Pattberg, and Widerberg, provides an innovative institutional mapping of the climate-energy nexus as a whole by applying the coherence dimensions of the analytical framework. The mapping presents a newly built database and novel methodology to identify intergovernmental or transnational institutions with a direct objective to steer society in various directions in terms of energy and climate change. It uses a 'governance triangle' developed by Abbott and Snidal (2009) to categorize the various institutions depending on their constitutive members and other institutional characteristics such as governance functions and thematic focus. The results show nearly 110 institutions in the climate-energy nexus, comprising more than 13,000 members, including public, private, and civil society actors, of which more than 8,000 are cities. The chapter also offers a network analysis of the nexus, demonstrating the tight web of institutions connected by primarily public actors and the centrality of the European Union and a few countries, including the United States and the United Kingdom, within that web.

Chapter 3 also provides the empirical basis for the further analyses in the subsequent chapters: it zooms in on the institutional constellations of the three subfields of renewable energy, fossil fuel subsidy reform, and carbon pricing. Similar to the analysis of the nexus as a whole, the chapter highlights for each of these subfields the constitutive characteristics of the respective institutions and the governance functions they perform. The results are first insights into the differences, commonalities, and varying degrees of complexity across the three institutional complexes. In sum, the chapter provides the first step in creating a knowledge base to guide actors that aim to navigate the institutionally complex global climate and energy governance system.

1.4.2 Coherence and Management in the Climate-Energy Nexus

Part II of the book zooms in on the three subfields and applies the analytical framework's dimensions of coherence and management.

Chapter 4 by Sanderink examines coherence and management for the renewable energy subfield. After introducing the central role of renewables in the global energy transition, the chapter presents the wide array of forty-six public, private, and public–private institutions that currently govern the subfield. These include intergovernmental organizations such as the International Energy Agency (IEA) and IRENA, but also private initiatives and multi-stakeholder partnerships such as RE100 and REN21. These various institutions not only differ in terms of their institutional characteristics but also with regard to the energy sources and technologies they cover, the governance functions they perform, and the energy-related objectives they prioritize.

Following this analysis of the meso level, the chapter concentrates on three multi-stakeholder partnerships to determine micro-level coherence and identify inter-institutional management attempts: the Renewable Energy and Energy Efficiency Partnership (REEEP), REN21, and Sustainable Energy for All (SEforALL). This assessment is based on a thorough analysis of the institutional characteristics of the selected partnerships, a qualitative review of their official documents, and semi-structured interviews with the experts involved. The result is a detailed overview of the institutional overlaps and differences, various interaction mechanisms, and several management attempts between and beyond these institutions. Based thereon, Chapter 4 concludes to what extent renewable energy subfield can be characterized in terms of categories such as division of labour, synergy, coexistence, or competition.

In Chapter 5, Verkuijl and van Asselt examine coherence and management across the institutional complex of fossil fuel subsidy reform. By reducing carbon emissions and enabling the redirection of public funds into areas such as health, education, and renewable energy, the reform of subsidies for the production and consumption of fossil fuels can support the goals of the Paris Agreement as well as a range of broader sustainable development objectives. However, although the benefits of their reform have been widely acknowledged, fossil fuel subsidies persist in both developed and developing economies. Most scholars addressing this puzzle have done so primarily through the lens of domestic politics.

Chapter 5, by contrast, considers how various international institutions are approaching FFSR governance. The authors briefly introduce the rationale for FFSR, before analyzing the coherence of the institutional complex at the meso level. This includes the possible emergence of a core norm of FFSR, membership distribution, and the governance functions carried out by the various international institutions active in this area. To further evaluate the degree of coherence in this field, the chapter zooms in on the micro level. Concretely, the authors examine a subset of three international clubs whose FFSR activities are among the most prominent globally: the Group of 20 (G20), the Asia-Pacific Economic Cooperation (APEC), and the Friends of Fossil Fuel Subsidy Reform. They first introduce the FFSR activities undertaken by each of these three institutions, and then, drawing on interviews, consider the interlinkages between these activities, as well as efforts to manage them. The chapter concludes by considering implications of the findings for the future management of FFSR governance and the complexity thereof.

Chapter 6 by Skovgaard and Canavan examines the third case study of the complex of institutions promoting carbon pricing. Carbon pricing – in the shape of carbon taxes, emissions trading, and offsets – places a price on the emission of carbon and mainly targets emissions from energy consumption. Advocates of carbon pricing have promoted it as the solution, or at least as one of the key tools to combat climate change. Even if one does not subscribe to these claims, carbon pricing can provide significant emissions reductions as well as fiscal revenue. Previous studies of carbon pricing have mainly focused on economic aspects and on single cases of pricing efforts at national, provincial, and European Union levels. This chapter contributes to the literature by focusing on the neglected issue of what the governance of carbon pricing looks like on the international level. It does so by mapping the thirteen international institutions that constitute the carbon-pricing subfield and by assessing the degree of cross-institutional coherence, i.e.

identifying to what extent the institutions can be organized into specific clusters. The examined carbon pricing institutions are public, private, and hybrid in nature. They overlap in terms of membership, geographical scope, and governance functions and to some degree differ in their interpretation of the core norm of carbon pricing.

Moreover, Chapter 6 focuses on the interlinkages between, on the one hand, the UNFCCC and, on the other hand, three World Bank-based institutions: the Carbon Pricing Leadership Coalition, the Networked Carbon Markets, and the Partnership for Market Readiness. These interlinkages, the (cognitive, normative, or behavioural) mechanisms through which they take place, and the attempts to manage them are studied through key informant interviews and official documents. The chapter finally discusses to which extent the subfield of carbon pricing can be characterized in terms of synergy, division of labour, coexistence, coordination, conflict, or competition.

1.4.3 Legitimacy and Effectiveness in the Climate-Energy Nexus

Part III examines the consequences of the degree of institutional complexity that we identified in the previous part. Two chapters will examine how the legitimacy and effectiveness of institutions can be assessed against the backdrop of the normative, functional, and membership-related coherence of the subfields in which these institutions are embedded.

Chapter 7 applies the evaluative theme of legitimacy that we introduced in Chapter 2. Nasiritousi and Verhaegen examine how stakeholders assess a set of institutions within the renewable energy subfield along nine dimensions of legitimacy. Given the proliferation of institutions with overlapping mandates, institutions rely on favourable legitimacy perceptions by a range of stakeholders in order to attract members and resources. By focusing on those aspects of legitimacy that the institutions themselves can affect, the chapter examines how assessments of institutional qualities differ between a diverse set of stakeholder groups.

Specifically, the chapter analyzes results from an expert survey among energy and climate stakeholders, including governmental and nongovernmental stakeholders from different world regions. Respondents were asked to assess five climate and energy governance institutions that exhibit different but overlapping mandates and membership: the IEA, CEM, IRENA, REN21, and UNFCCC. Through this systematic and comparative mapping of stakeholders' legitimacy assessments, the chapter offers novel insights into legitimacy under institutional complexity and concludes by discussing implications for institutions' legitimation strategies.

In Chapter 8, Nasiritousi, Sanderink, Skovgaard, van Asselt, Verkuijl, and Widerberg ask: How do institutional linkages affect the performance and the effectiveness of individual institutions and how can performance and effectiveness be strengthened in the institutionally complex nexus of global climate and energy governance? The chapter addresses these questions for the three subfields studied in this book: renewable energy, fossil fuel subsidy reform, and carbon markets. It starts off with discussing the concept of effectiveness and highlighting the challenges to analyzing the effectiveness of institutions, especially when they have overlapping mandates and are interlinked. In order to respond to these challenges, our research relies on a two-track approach, integrating assessments by researchers and interviews with key stakeholders.

By examining the outputs, outcomes, and impacts of the three subfields, the chapter shows both the advantages and the disadvantages of institutional complexity in the climate-energy nexus for achieving effectiveness. Through its crosscutting perspective, the study identifies a set of management options and discusses opportunities and barriers to reform the nexus. It concludes by outlining implications for overcoming some of the downsides of institutional complexity.

The conclusions of this book in Chapter 9 provide a critical review of the conceptual, empirical, and theoretical contributions in this volume. First, we summarize the major novelties and findings, systematically contrasting the empirical results for coherence, management, legitimacy, and effectiveness across the subfields of the climate-energy nexus. Second, the chapter explores to which extent the results across the institutional complexes for renewable energy, FFSR, and carbon pricing can be explained or understood by the problem-structural approach we introduced previously – i.e. by the nature of the problem to be regulated and its position within the climate-energy nexus – or by alternative approaches. Third, we discuss potential policy recommendations that could be derived from the findings. Finally, the chapter suggests a future research agenda on the governance of the climate-energy nexus.

1.5 References

Abbott, K. W. and Snidal, D. 2009. The Governance Triangle: Regulatory Standards Institutions and the Shadow of the State. In *The Politics of Global Regulation*, edited by W. Mattli and N. Woods. Princeton: Princeton University Press.

Andresen, S. and Hey, E. 2005. The Effectiveness and Legitimacy of International Environmental Institutions. *International Environmental Agreements: Politics, Law and Economics* 5(3), 211–226.

Bäckstrand, K. 2008. Accountability of Networked Climate Governance: The Rise of Transnational Climate Partnerships. *Global Environmental Politics* 8(3), 74–102.

Bäckstrand, K., Kuyper, J. W., Linnér, B.-O., and Lövbrand, E. 2017. Non-State Actors in Global Climate Governance: From Copenhagen to Paris and Beyond. *Environmental Politics* 26(4), 561–579.

Bäckstrand, K., Zelli, F., and Schleifer, P. 2018. Legitimacy and Accountability in Polycentric Climate. In *Governance Governing Climate Change: Polycentricity in Action?* edited by A. Jordan, D. Huitema, H. van Asselt, and J. Foster. Cambridge: Cambridge University Press.

Barnsley, I. and Ahn, S.-J. 2014. *Mapping Multilateral Collaboration on Low-Carbon Energy Technologies*. Paris: International Energy Agency.

Bazilian, M., Hobbs, B. F., Blyth, W., MacGill, I., and Howells, M. 2011. Interactions between Energy Security and Climate Change: A Focus on Developing Countries. *Energy Policy* 39(6), 3750–3756.

Biermann, F., Pattberg, P., van Asselt, H., and Zelli, F. 2009. The Fragmentation of Global Governance Architectures: A Framework for Analysis. *Global Environmental Politics* 9(4), 14–40.

Boas, L., Biermann, F., and Kanie, N. 2016. Cross-Sectoral Strategies in Global Sustainability Governance: Towards a Nexus Approach. *International Environmental Agreements: Politics, Law, and Economics* 16(3), 449–464.

Bradshaw, M. J. 2014. *Global Energy Dilemmas: Energy Security, Globalization, and Climate Change*. Cambridge: Polity Press.

Buchanan, A. and Keohane, R. 2006. The Legitimacy of Global Governance Institutions. *Ethics and International Affairs* 20(4), 405–437.

Bulkeley, H., Andonova, L. B., Betsill, M. M., Compagnon, D., Hale, T., Hoffmann, M. J., Newell, P., Paterson, M., Roger, C., and VanDeveer, S. D. 2014. *Transnational Climate Change Governance*. New York: Cambridge University Press.

Chambers, W. B. 2008. *Interlinkages and the Effectiveness of Multilateral Environmental Agreements*. Tokyo: United Nations University Press.

Charnovitz, S. 2003. World Trade Organization and Law Enforcement. *Journal of World Trade* 37(5), 817–837.

Cherp, A., Jewell, J., and Goldthau, A. 2011. Governing Global Energy: Systems, Transitions, Complexity. *Global Policy* 2(1), 75–88.

Clancy, J., Skutsch, M., and Batchelor, S. 2002. The Gender-Energy-Poverty Nexus: Finding the Energy to Address Gender Concerns in Development. DFID Project CNTR998521.

Dubash, N. K. and Florini, A. 2011. Mapping Global Energy Governance. *Global Policy* 2(1), 6–18.

Dyer, H. C. 2014. Climate Anarchy: Creative Disorder in World Politics. *International Political Sociology* 8(2), 182–200.

Eberlein, B., Abbott, K. W., Black, J., Meidinger, E., and Wood, S. 2014. Transnational Business Governance Interactions: Conceptualization and Framework for Analysis. *Regulation & Governance* 8(1), 1–21.

Falkner, R. 2014. Global Environmental Politics and Energy: Mapping the Research Agenda. *Energy Research and Social Science* 1, 188–197.

Falkner, R. 2018. Climate Change, International Political Economy and Global Energy Policy. In *Handbook of the International Political Economy of Energy and Natural Resources*, edited by A. Goldthau, M. F. Keating, and C. Kuzemko. Cheltenham, UK: Edward Elgar Publishing.

Food and Agriculture Organization of the United Nations. 2014. *The Water-Energy-Food Nexus: A New Approach in Support of Food Security and Sustainable Agriculture*. Rome: Food and Agriculture Organization of the United Nations.

Gehring, T. and Oberthur, S. 2009. The Causal Mechanisms of Interaction between International Institutions. *European Journal of International Relations* 15(1), 125–156.

Goldthau, A., Keating, M. F., and Kuzemko, C. 2018. *Handbook of the International Political Economy of Energy and Natural Resources*. Cheltenham, UK: Edward Elgar Publishing.

Goldthau, A. and Witte, J. M. 2010. *Global Energy Governance: The New Rules of the Game*. Washington, DC: Brookings Institution Press.

Heubaum, H. and Biermann, F. 2015. Integrating Global Energy and Climate Governance: The Changing Role of the International Energy Agency. *Energy Policy* 87, 229–239.

Hjerpe, M. and Nasiritousi, N. 2015. Views on Alternative Forums for Effectively Tackling Climate Change. *Nature Climate Change* 5(9), 864–867.

Hoff, H. 2011. Understanding the Nexus. Background Paper for the Bonn 2011 Conference: The Water, Energy and Food Security Nexus. Stockholm: Stockholm Environment Institute.

Hoffmann, M. J. 2011. *Climate Governance at the Crossroads. A Global Response after Kyoto*. Oxford: Oxford University Press.

IEA (International Energy Agency). 2018. *World Energy Outlook 2018*. Paris: IEA.

Iguchi, M., Ehara, T., Yamazaki, E., Tasaki, T., and Abe, N. 2014. Ending the Double Burden of Malnutrition: Addressing the Food and Health Nexus in the Sustainable Development Goals. POST2015/UNU-IAS Policy Brief #8.

Jacquemont, F. and Caparrós, A. 2002. The Convention on Biological Diversity and the Climate Change Convention 10 Years after Rio: Towards a Synergy of the Two Regimes? *Review of European Community and International Environmental Law* 11(2), 169–180.

Jewell, J., McCollum, D., Emmerling, J., Bertram, C., Gernaat, D. E. H. J., Krey, V., Paroussos, L., Berger, L., Fragkiadakis, K., Keppo, I., Saadi, N., Tavoni, M., van Vuuren, D., Vinichenko, V. and Riahi, K. 2018. Limited Emission Reductions from Fuel Subsidy Removal Except in Energy-Exporting Regions. *Nature* 554, 229–233.

Jordan, A. J., Huitema, D., Hildén, M., van Asselt, H., Rayner, T. J., Schoenefeld, J. J., and Boasson, E. L. 2015. Emergence of Polycentric Climate Governance and Its Future Prospects. *Nature Climate Change* 5(11), 977–982.

Karlsson-Vinkhuyzen, S. I. 2010. The United Nations and Global Energy Governance: Past Challenges, Future Choices. *Global Change, Peace & Security* 22(2), 175–195.

Karlsson-Vinkhuyzen, S. I. and McGee, J. 2013. Legitimacy in an Era of Fragmentation: The Case of Global Climate Governance. *Global Environmental Politics* 13(3), 56–78.

Keohane, R. O. and Victor, D. G. 2011. The Regime Complex for Climate Change. *Perspectives on Politics* 9(1), 7–23.

Lesage, D. and Van de Graaf, T. 2016. *Global Energy Governance in a Multipolar World*. New York: Routledge.

Métivier, C., Postic, S., Alberola, E., and Vinnakota, M. 2018. Global Carbon Account 2018. Paris: Institute for Climate Economics, I4CE.

Moncel, R. and van Asselt, H. 2012. All Hands on Deck! Mobilizing Climate Change Action beyond the UNFCCC. *Review of European Community and International Environmental Law* 21(3), 163–176.

Munaretto, S. and Witmer, M. 2017. *Water-Land-Energy-Food-Climate Nexus: Policies and Policy Coherence at European and International Scale*. Hague: Netherlands Environmental Assessment Agency.

Nasiritousi, N. 2017. Fossil Fuel Emitters and Climate Change: Unpacking the Governance Activities of Large Oil and Gas Companies. *Environmental Politics* 26(4), 621–647.

Oberthür, S. 2009. Interplay Management: Enhancing Environmental Policy Integration Among International Institutions. *International Environmental Agreements: Politics, Law and Economics* 9(4), 371.

Oberthür, S. and Gehring, T. (Eds.). 2006. *Institutional Interaction in Global Environmental Governance: Synergy and Conflict among International and EU Policies*. Cambridge, MA: MIT Press.

Oberthür, S. and Stokke, O. S. (Eds.). 2011. *Managing Institutional Complexity: Regime Interplay and Global Environmental Change*. Cambridge, MA: MIT Press.

Ostrom, E. 2010. Polycentric Systems for Coping with Collective Action and Global Environmental Change. *Global Environmental Change* 20(4), 550–557.

Pattberg, P. and Widerberg, O. 2015. Theorising Global Environmental Governance: Key Findings and Future Questions. *Millennium* 43(2), 684–705.

Postic, S. and Métivier, C. 2019. *Global Carbon Account 2018*. Paris: Institute for Climate Economics, I4CE.

Röhrkasten, S. 2018. Are We on the Cusp of a Global Renewable Energy Transition? In *Oxford Handbook of Energy and Society*, edited by D. J. Davidson and M. Gross. Oxford: Oxford University Press.

Sanderink, L., Widerberg, O., Kristensen, K., and Pattberg, P. 2017. *Mapping the Institutional Architecture of the Climate-Energy Nexus IVM Report (R-17/04)*. Amsterdam: Institute for Environmental Studies, Vrije Universiteit Amsterdam.

Sovacool, B. K. 2011. An International Comparison of Four Polycentric Approaches to Climate and Energy Governance. *Energy Policy* 39(6), 3832–3844.

Sovacool, B. K. 2013. Energy Policy and Climate Change. In *The Handbook of Global Climate and Environment Policy*, edited by R. Falkner. Malden, MA: Wiley-Blackwell.

Sovacool, B. K. and Florini, A. 2012. Examining the Complications of Global Energy Governance. *Journal of Energy & Natural Resources Law* 30(3), 235–263.

Sovacool, B. K. and Van de Graaf, T. 2018. Building or Stumbling Blocks? Assessing the Performance of Polycentric Energy and Climate Governance Networks. *Energy Policy* 118, 317–324.

Stefanski, R. 2016. Into the Mire: A Closer Look at Fossil Fuel Subsidies. *SPP Research Papers* 9(10).

Terton, A., Gass, P., Merrill, L., Wagner, A., and Meyer, E. 2015. *Fiscal Instruments in INDCs: How Countries are Looking to Fiscal Policies to Support INDC Implementation*. Geneva: International Institute for Sustainable Development.

UNEP (United Nations Environment Programme) 2018. *Emissions Gap Report 2018*. Nairobi: UNEP.

van Asselt, H. 2014. *The Fragmentation of Global Climate Governance: Consequences and Management of Regime Interactions*. Cheltenham, UK: Edward Elgar.

van Asselt, H. and Zelli, F. 2014. Connect the Dots: Managing the Fragmentation of Global Climate Governance. *Environmental Economics and Policy Studies* 16(2), 137–155.

van Asselt, H. and Zelli, F. 2018. Polycentric Governing by and beyond the UNFCCC. In *Governance Governing Climate Change: Polycentricity in Action?* edited by A. Jordan, D. Huitema, H. van Asselt and J. Foster. Cambridge: Cambridge University Press.

Van de Graaf, T. 2013. *The Politics and Institutions of Global Energy Governance: Houndmills*. Basingstoke, Hampshire: Palgrave Macmillan.

Van de Graaf, T. and Blondeel, M. 2018. Fossil Fuel Subsidy Reform: An International Norm Perspective. In *The Politics of Fossil Fuel Subsidies and Their Reform*, edited by J. Skovgaard and H. van Asselt. Cambridge: Cambridge University Press.

Van de Graaf, T. and Colgan, J. 2016. Global Energy Governance: A Review and Research Agenda. *Palgrave Communications*, 2.

Van de Graaf, T. and Zelli, F. 2016. Actors, Institutions and Frames in Global Energy Politics. In *Palgrave Handbook of the International Political Economy of Energy*, edited by T. Van de Graaf, B. K. Sovacool, F. Kern, A. Ghosh, and M. T. Klar. London: Palgrave Macmillan.

Vogler. 2018. Energy, Climate Change, and Global Governance: The 2015 Paris Agreement in Perspective. In *Oxford Handbook of Energy and Society*, edited by D. J. Davidson and M. Gross. Oxford: Oxford University Press.

Widerberg, O. and Pattberg, P. 2015. International Cooperative Initiatives in Global Climate Governance: Raising the Ambition Level or Delegitimizing the UNFCCC? *Global Policy* 6(1), 45–56.

Widerberg, O. and Stenson, D. E. 2013. *Climate Clubs and the UNFCCC*. Fores Study 3, Stockholm: Fores.

Wilson, J. D. 2015. Multilateral Organizations and the Limits to International Energy Cooperation. *New Political Economy* 20(1), 85–106.

Young, O. R. 1996. Institutional Linkages in International Society: Polar Perspectives. *Global Governance* 1(2), 1–24.

Young, O. R., Chambers W. B., Kim, J. A., and ten Have, C. (Eds.). 2008. *Institutional Interplay: Biosafety and Trade*. Tokyo: United Nations University Press.

Zelli, F. 2011. The Fragmentation of the Global Climate Governance Architecture. *WIREs: Climate Change* 2(2), 255–270.

Zelli, F. and van Asselt, H. 2013. The Institutional Fragmentation of Global Environmental Governance: Causes, Consequences, and Responses. *Global Environmental Politics* 13(3), 1–13.

Zelli, F., Pattberg, P., Stephan, H., and van Asselt, H. 2013. Global Climate Governance and Energy Choices. In *The Handbook of Global Energy Policy*, edited by A. Goldthau. Chichester: Wiley-Blackwell.

Part I

Mapping the Climate-Energy Nexus

2

Analytical Framework

Assessing Coherence, Management, Legitimacy, and Effectiveness

FARIBORZ ZELLI, NAGHMEH NASIRITOUSI, KARIN BÄCKSTRAND, PHILIPP PATTBERG, LISA SANDERINK, JAKOB SKOVGAARD, HARRO VAN ASSELT, AND OSCAR WIDERBERG

2.1 Introduction: Approaching the Climate-Energy Nexus

How can we analyze the complexity of the climate-energy nexus in a systematic way that is informative for both scholars and practitioners? Over the past roughly twenty years, scholars discussed a series of approaches to make sense of the growing institutional complexity in international relations. They mostly did so for what we call the micro level in this volume, i.e. for interactions between individual institutions, and in particular between intergovernmental institutions (e.g. Young 1996; Stokke 2001; Gehring and Oberthür 2009). Some of the more recent scholarship completed these approaches by targeting relations among transnational arrangements (e.g. Eberlein et al. 2014; Tosun et al. 2016). The merits of these different perspectives notwithstanding, they led to certain terminological squabbles – with preferences for different concepts such as polycentricity, polyarchy, or complexity – while mostly neglecting overlaps between a larger set of institutions.

As argued in Chapter 1, what is missing is a pragmatic approach (1) that combines these literatures and converges their often similar categories for (a) a nexus among two major policy fields, climate and energy, and (b) different analytical levels within such a nexus, and (2) that also takes into account consequences of institutional interactions and complexity. In the next pages we propose such a careful compromise of dimensions. This approach will guide not only our analyses of the institutional coherence and management of the climate-energy nexus as a whole (Chapter 3) and of three selected subfields within this nexus (Chapters 4–6), but also our examinations of respective consequences for legitimacy and effectiveness (Chapters 7 and 8).

The framework leaves space for the case studies in this volume to unravel additional dimensions that are suitable to the respective policy field or subfield and research question they are addressing. We hope that this multi-dimensional framework with its mix of novelty and flexibility will also be attractive for future qualitative assessments of linkages between private, public, and hybrid institutions in other global governance fields.

The chapter will proceed as follows. The next section briefly lays the conceptual ground by disentangling major concepts (nexus, institutions, and types thereof) and analytical levels that we put under scrutiny in this volume. This is followed by the main Section 2.3, which establishes the core evaluative themes of our analytical framework – coherence, management, legitimacy, and effectiveness – as well as dimensions to assess them for different subfields and analytical levels. The concluding section summarizes our approach and looks ahead to its application in the remainder of this volume.

2.2 Conceptual Framework

2.2.1 Core Concepts

Before starting with our actual framework, we briefly introduce key terms and outline how we understand them in this chapter and the remainder of the volume. We define *global governance* as 'all coexisting forms of collective steering of social affairs, by public and private actors, that directly or in their repercussions, transcend national frontiers' (Zelli 2018). *International institutions*, be they intergovernmental or transnational, are one major instrument to provide such collective steering. We follow Robert Keohane (1989, 3) who defined institutions as 'persistent and connected sets of rules (formal and informal) that prescribe behavioural roles, constrain activity, and shape expectations'. This definition treats institutions as the most generic term for forms of collaboration at the international level, since it neither specifies the degree of persistence or connectedness nor the types of members or functions involved.

This leaves space for a set of types of institutions. These include *organizations*, i.e. institutions with 'purposive entities . . . capable of monitoring activity and of reacting to it' (Keohane 1989, 3) and *regimes*, i.e. 'institutions with explicit rules . . . that pertain to particular sets of issues in international relations' (Keohane 1993, 28). In addition to these traditional types of international institutions, there are more loosely coupled types, which include *initiatives*, i.e. connected rule systems with a relatively low degree of persistency (meaning that they may be short-lived or very malleable), and *networks*, i.e. institutions with low degrees of both persistency and connectedness.

All institutions that address a specific policy field (or, as a synonym, domain) of international relations form the *institutional complex* for that field – with *institutional architecture* as a term we use synonymously on occasions. Wherever two institutional complexes overlap, forming a set of institutions that address two different policy fields in parallel, we speak of an *institutional nexus*. The delineation of a policy field is hence key for the inclusion or exclusion of institutions in a complex or nexus. This delineation, we argue, has a necessarily constructivist or subjective aspect, i.e. it depends on how certain observers or participants frame an issue in question. We provided our own understandings of climate change and energy as well of the three subfields we chose in Chapter 1, but of course acknowledge that this is but one possible way of delineating these. Being mindful of this perceptional or framing element, our analyses will rely not only on rationalist but also on constructivist evaluative themes, as we further clarify in Section 3.

2.2.2 Analytical Levels

One core contribution of our analytical approach is the distinction of, and positioning within, three levels of analysis that we often find mixed up in the emerging literature on institutional complexity: the macro level, which captures the overarching institutional fragmentation for a given field or domain of global governance and its respective complex; the meso level, which zooms into specific subfields of such a domain and the institutional complexes thereof; and the micro level, which exhibits interlinkages among two individual institutions on specific questions (cf. Biermann and Kim, forthcoming).

With regard to the thematic scope of our volume, this distinction of analytical levels translates into: at the macro level, the climate-energy nexus as a whole; at the meso level, the three subfields of this nexus that we put under scrutiny in the following chapters, i.e. carbon pricing, fossil fuel subsidies, and renewable energies; and at the micro level, specific interlinkages within each of these subfields that we studied, e.g. between the Group of 20, the Asia Pacific Economic Cooperation, and the Friends of Fossil Fuel Subsidy Reform in Chapter 5. Figure 2.1 illustrates how these three levels of institutional complexity relate to each other.

It is particularly the meso level that so far has gained only little attention by the literature on institutional complexity. It is here where this volume fills an important research gap. In the next section, we introduce a set of evaluative themes that help us to analyze interlinkages at the micro level in their relevance for the meso level, i.e. for a subfield of the climate-energy nexus. These four evaluative themes are:

Coherence and Management: to characterize sub-themes of the nexus (meso-level), and specific institutional linkages therein (micro level), according to their normative, functional, and membership-based relations and the deliberate attempts to regulate these relations;

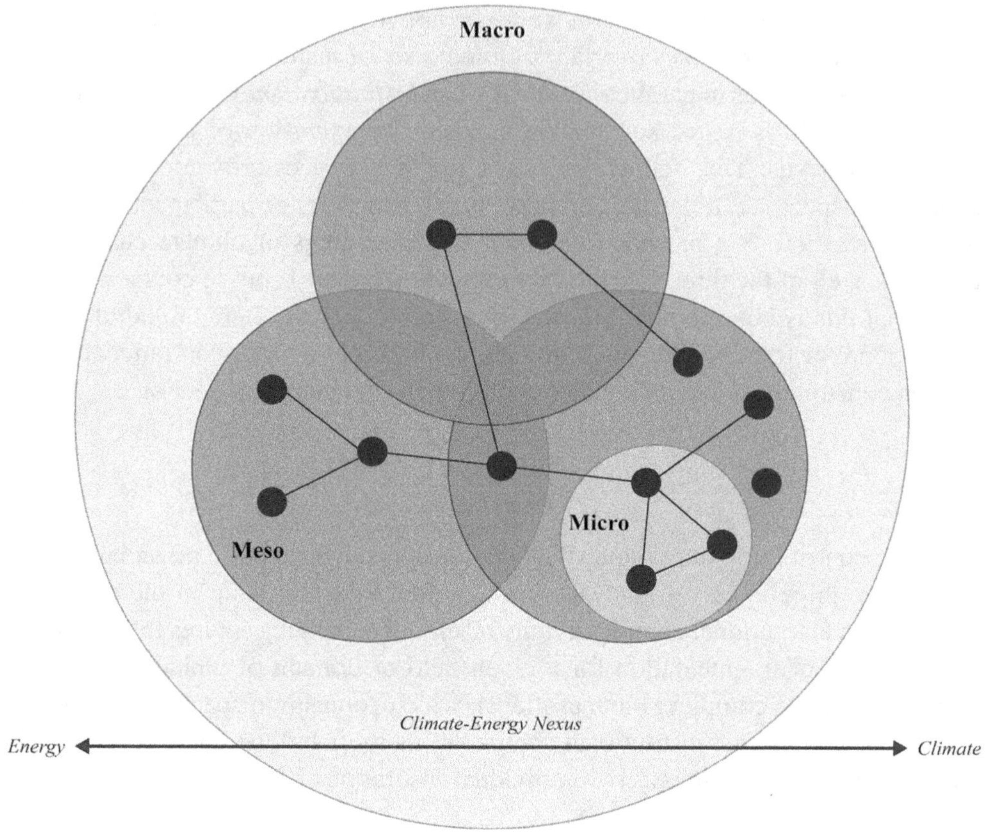

Figure 2.1 Analytical levels of the climate-energy nexus.

Legitimacy and *Effectiveness*: to characterize respective consequences of micro-linkages for the subfields to which they belong.

2.3 Evaluative Themes

In ontological terms, we follow a dual approach to give us a more comprehensive picture when analyzing the climate-energy nexus. Concretely, the four evaluative themes that we introduce in the following help us to capture both rationalist and constructivist aspects at different levels. For the first two – coherence and management – we apply a rationalist or objectivist approach by examining factual dimensions, e.g. membership patterns or concrete efforts taken to manage an institutional interlinkage. For the themes of legitimacy and effectiveness, by contrast, we bring in a sociological approach, i.e. we identify respective perceptions and expectations of actors attributed to the relationships among institutions within a nexus.

2.3.1 Coherence

How then can we qualify institutional interactions at the micro level and their relevance for the meso level, i.e. for a specific subfield within a nexus of two policy fields, in a feasible and qualitative way that allows for comparisons across such interactions? When can we speak of rather conflictive or synergistic cross-institutional relationships within a policy subfield, and on what grounds? We define coherence in institutional complexes as the harmony of certain institutional features to one another or to an overarching purpose. To characterize such institutional coherence in a policy subfield such as carbon pricing, we suggest two consecutive steps that build on the same dimensions for assessment (Box 2.1).

(1) The first step of analyzing meso-level coherence provides input on three analytical dimensions – core norm, membership, governance functions – that will guide the subsequent two steps of assessing institutional linkages, and their management, within the respective policy subfield. This step is largely descriptive in nature, since it assesses the three dimensions according to their distribution across the climate-energy nexus.

(1a) The first and most central dimension, *core norm* of a policy field, depends on the delineation and essential questions of the subfield under scrutiny. We follow a sociological understanding of norms here, not a narrower legal one, as ‘shared expectations about appropriate behaviour held by a community of actors’ (Finnemore 1996, 22). For the realm of global governance, this implies a shared narrative

Box 2.1
Dimensions of Coherence

(1) *Meso-level Coherence*, i.e. the level of coherence in the subfield as a whole, to be assessed along three major dimensions
   a. *Core norm*, i.e. the overarching expectation of behaviour that characterizes the subfield, and how this expectation has been interpreted in institutional practice;
   b. *Membership*: public, private, or public–private, across institutions in the subfield;
   c. *Governance functions* addressed by different institutions and their members;

(2) *Micro-level Coherence*, i.e. the level of coherence in specific interactions of institutions within the respective subfield, in terms of the same three dimensions and underlying mechanisms
   a. *Core norm* (and each institution’s interpretation of it);
   b. *Membership*;
   c. *Governance functions*;
   d. *Mechanisms* of interactions among institutions.

about the conditions and objectives that inform individual institutions and even entire institutional complexes (or architectures) as a whole (Ruggie 1998). In other words, building on Conca (2006) and Zelli et al. (2013),we hold that such complexes are embedded in a broader normative context. For instance, the core norm of the global climate governance complex would be to 'prevent dangerous anthropogenic interference with the climate system' as laid out in Article 2 of the United Nations Framework Convention on Climate Change.

The first task for our researchers is to identify such a core norm for the policy subfield they put under scrutiny, e.g. from the preambles of leading institutions in the respective subfield. Following this, they need to assess whether there is a certain convergence among interpretations of this norm across different institutions and their linkages, or whether these institutions follow different interpretations or variants of the norm.

Importantly, the identification of a core norm is closely related to the delineation of a subfield, or rather its institutional complex – since the core norm provides the substance or goal that unites the institutions governing the field in question. As discussed in Chapter 1, this implies that the delineation of a governance problem or institutional complex has an element of construction to it. In formulating the core norms for each of the three subfields (see Chapters 4, 5, and 6), the authors make a subjective choice as researchers about what counts as a sub-structure within the climate-energy nexus.

The construction of the core norm provides a crucial link between the evaluative theme of coherence and the other three themes of this framework. Management can be seen as an active attempt to enhance normative coherence across institutions by creating a common understanding of a problem and working toward a joint goal with different institutional efforts. And legitimacy and effectiveness are conceptualized sociologically in the further course of this chapter, i.e. in terms of assessments or perceptions, which directly link back to the core norm underlying an institutional complex: institutions may be assessed according to whether they have the right to govern toward that norm (legitimacy) and whether they deliver on that norm (effectiveness).

The centrality of the (constructed) core norm in our framework also has consequences for how we understand normative coherence: there is a bias toward convergence on the core norm as a positive or at least essential ideal for a subfield in question, since too many strong deviations from that norm would undermine the identification of the subfield in the first place.

(1b+1c) By contrast, we treat the two ensuing dimensions – *membership* and *governance functions*, i.e. coherence in their case is not marked by convergence but by balanced distribution and synergy. These two dimensions will also be used

Box 2.2
Coherence Dimension of Membership – Types

- Public
  - International organizations / bureaucracies (intergovernmental or transnational);
  - States (could be further distinguished: governments, specific ministries, agencies);
  - Sub-national regions / provinces / states;
  - Cities;
- Civil Society Organizations
  - Nongovernmental organizations (could be further distinguished, e.g. advocacy, watchdog, disruptive);
  - Research / Experts (could be further distinguished, e.g. certifiers, academic observers);
- Firm
  - Business;
  - Investors.

in our analysis of the macro-level coherence of the entire climate-energy nexus in Chapter 3.

Boxes 2.2 and 2.3 display our typologies for membership and governance functions that institutional complexes and their components may exhibit in different mixes. We build the respective types on Eberlein et al. (2014), Tosun et al. (2016) as well as Abbott (2012) and Abbott and Snidal (2009a; 2009b) with their governance triangle and distinction of functions (see also Chapter 3 for a more detailed introduction of both typologies).

Coherence depends on the extent to which the different types of members (or audiences) and governance functions are covered across an institutional complex, and on the extent to which this distribution or coverage is taking place in a synergistic way. Thus, an institutional complex exhibits low coherence if, for instance, operational activities are hardly performed by institutions and major public or private actors are not members. Importantly, certain activities can pertain to various governance functions, e.g. a report on financing may fall under both financing and information and networking.

With regard to synergy, the activities depicted in Box 2.3 can interact in different ways. Information-sharing and networking do not inherently have the same potential for competition or conflict as the setting of standards and commitments, since the adding of another standard- or commitment-setting institution to a subfield often undermines the existing standard- or commitment-setting

Box 2.3
Coherence Dimension of Governance Functions – Types

- Standards and commitments, e.g.:
  - Rule-making processes;
  - Mandatory commitments;
  - Voluntary commitments (e.g. voluntary pollution reduction schemes);
  - Communiqués;
  - Schemes for implementation and enforcement;
  - Standards for monitoring, reporting and verification.
- Operational activities, e.g.:
  - Technology research and development;
  - Pilot projects;
  - Demonstration and deployment activities;
  - Technical assistance and capacity building;
  - Best practices;
  - Implementation of commitments;
  - Enforcement activities.
- Information and networking, e.g.:
  - Monitoring, reporting, and verification;
  - Evaluation activities (e.g. peer reviews);
  - Behind-the-scenes diplomacy;
  - Collecting and publishing information;
  - Technical consulting, training, and information events & services to build capacity, share knowledge, and to support local governments;
- Financing (i.e. operational activities that particularly involve financing schemes), e.g.:
  - Project-funding;
  - Conditions in structural adjustment programmes;
  - Aid.

institutions. If, for instance, one institution sets standards for offsets, and another institution starts to introduce a different set of standards for offsets, competition or conflict is likely, inter alia, because of forum or 'standard' shopping by public and private actors (Alter and Meunier 2009). Information-sharing and networking are different from standard-setting in that the information-sharing and networking of one institution is likely to add to the information-sharing and networking of another institution, provided that their activities are connected.

(2) Our in-depth analysis of *micro-level coherence* will equally be based on the three aforementioned dimensions, i.e. we suggest going beyond legal consistency

toward assessing *normative, membership-based, and functional coherence* among specific institutions in a policy subfield.

(2a+b+c) This threefold understanding implies that complete coherence, like complete conflict, is rather an ideal type – since we will rarely find them to the same degree across all of these three dimensions. Two institutions that address a particular energy policy issue such as fossil fuel subsidy reform may, for instance, be synergistic or at least non-conflictive in terms of the core norm they adhere to, e.g. the need to phase out respective subsidies, but still may invade each other's turfs and compete over governance functions they seek to serve.

When it comes to weighing the coherence of the three core dimensions against each other, we suggest a simple hierarchy, holding that normative relationships between institutions in a policy subfield are most important, followed by functional ones, with membership carrying the least weight. If two institutions move into very different directions on major norms or governance functions (e.g. egalitarian versus efficiency goals on carbon pricing, or rivalling ideas for standards or evaluation), this constitutes a major conflict that cannot be balanced on the level of membership, which is rather formal in nature.

The researcher therefore has to establish whether the core norms and governance functions between two institutions turn out rather synergistic or conflicting. This implies a careful qualitative analysis about the *essence and direction* of each institutional approach, i.e. a combination of normative and functional assessment. It is not sufficient to check whether certain governance functions are complimentary, duplicated, or not served at all.

For instance, one might find that intergovernmental institutions are mostly responsible for the formulation of standards or targets, while private institutions largely take over information and networking tasks. At first glance, this may suggest a sensible division of labour at the meso level. However, a closer look at the micro level, i.e. at the actual essence of the activities within and across selected institutions, might reveal that public institutions adhere to different norms and pursue different goals and ambitions (e.g. command-and-control and trade-restrictive) than private ones (e.g. bottom-up and market-liberal) – and that this leads to a disconnect among the purposes of institutions.

The coherence in membership, finally, is closely connected to governance functions. The focus here is on how different actors in a given part of the nexus have these functions covered in a coherent manner, i.e. which enabling or restricting roles are played to fulfil certain governance functions.

(2d) Importantly, our assessment does not stop at contrasting individual institutions in terms of their functions, membership, and norms. As mentioned earlier in this chapter, the micro level of complexity for us is constituted by institutional interlinkages, i.e. relations between institutions along the dimensions we discussed. For a fuller picture on coherence in a policy subfield this implies that: we need to scrutinize the actual exchange processes among institutions, not just compare their individual norms, memberships, and functions – and qualify these exchanges in terms of underlying causal mechanisms.

To characterize different *interaction mechanisms* that are of relevance in such exchanges across norms, functions, and members, we rely on the literature on causal mechanisms in institutional interlinkages (Stokke 2001, 10–23; Gehring and Oberthür 2006; Oberthür and Gehring 2006), but also on more recent typologies, e.g. Eberlein et al. (2014) and Tosun et al. (2016), and our own work, e.g. Zelli et al. (2013) and Van de Graaf and Zelli (2016).

Concretely, we distinguish interactions at the cognitive, normative, and behavioural levels. Box 2.4 displays our definitions and examples for these three types.

This selection shown in Box 2.4 is non-exhaustive and we are aware that other scholars added further mechanisms, e.g. legal or institutionalized interactions among institutions (Eberlein et al. 2014). We take out such legal interactions here and deal with them under the evaluative theme of management, since we see them as results of management attempts, not as processes of interaction.

Box 2.4

Coherence Dimension of Interaction Mechanisms – Types

- Cognitive = impact of knowledge and information in one institution on another one, e.g.:
  - Flow of information between institutions and their members;
  - Deeper cross-institutional learning, including on the execution of certain governance functions;
- Normative = impact of norms and rules of one institution on another one, e.g.:
  - Imitation of rules / practices;
  - Adaptation toward certain rules or practices;
  - Internal debates over rules of another institution;
- Behavioural = impact of the functional and strategic behaviour of members and other relevant actors in one institution on another one, e.g.:
  - interlocking memberships;
  - Pressure;
  - Shaming;
  - 'Monitoring' each other's performance;
  - Brand management;
  - Capacity-building, e.g. resource allocation.

For each of these mechanisms we will assess to what extent they affect the levels of functional, normative, and membership-based coherence among institutions in a policy subfield. As some of the examples in Box 2.4 suggest, certain mechanisms may well have more importance for certain dimensions than others. For instance, membership and certain governance functions play a key role for the behavioural mechanism, while norms are at the core of the normative mechanism.

2.3.2 Management

When we refer to the evaluative theme of management in this volume, we mean *micro-level management*, which for us describes any deliberate attempt to formally regulate the linkage between two or more institutions (Zelli 2010). This follows Stokke's understanding of interplay management as 'deliberate efforts by participants in tributary or recipient regimes to prevent, encourage, or shape the way one regime affects problem solving under another' (Stokke 2001, 11).

Management in this understanding is one possible form of intentionality in institutional linkages. Gehring and Faude (2014) distinguish further types of intentionality, especially the very creation of a functional overlap among two institutions in the first place, e.g. to challenge the regulatory dominance of one of them (cf. Schneider and Urpelainen 2013; Van de Graaf 2013) or to shift regulatory activities from one to the other (Braithwaite and Drahos 2000, 564–577; Helfer 2009).

This notwithstanding, we focus here on management instead of intentionality as a whole. One reason for our choice is that a wider perspective on intentionality would mix causal explanations into our assessment of the character of an interaction or sub-complex – and we intend to keep the former apart from the latter as far as this is analytically possible. Moreover, management has been more systematized in the literature than intentionality, giving us a set of tools at hand to clarify and characterize the state of management of an institutional linkage.

We roughly distinguish two dimensions by which we assess micro-level management for our three case studies (Box 2.5).

Box 2.5
Dimensions of Management

(1) Levels and agents of management;
(2) Consequences of management, i.e. any change or convergence on the coherence dimensions (norms, membership, governance functions).

Box 2.6
Management Dimension of Levels and Agency – Types

- Uncoordinatedly at national or regional levels;
- Unilaterally by one of the affected institutions;
- Jointly between the affected institutions; or
- By an overarching institution (e.g. the United Nations Environment Programme) or third institutions (e.g. the International Court of Justice) as mediator or dispute settler.

(1) Following Oberthür's and Stockke's influential typologies, management attempts can be distinguished according to the *levels* at which they are taken and the types of *agents* usually involved at a respective level (Oberthür 2009; Oberthür and Stokke 2011). This yields four types of *management levels and agents* (Box 2.6).

(2) In addition, management brings in a time dimension. It implies that, at least analytically, one may distinguish essential aspects of an inter-institutional relation – that we scrutinize under the evaluative theme of 'coherence' – from a reaction to this relation. Thus, a second dimension that we seek to assess for micro-level management is the *consequences* of the management process. Using the aforementioned coherence dimensions, we examine to what extent management efforts enhanced the functional, normative, and membership-based coherence or division of labour among institutions in a subfield. As for the assessment of mini-level coherence that we introduced earlier in this chapter, it is particularly crucial to assess the normative and functional consequences of the respective management attempts, e.g. whether or not they lead to more synergy and, more precisely, whether the management approach ultimately serves the functions and goals of one or both institutions.

To summarize this and the previous subsection, we have established a set of dimensions and associated types to analyse two of our evaluative themes (coherence at the meso and micro levels, and management at the micro level). These steps will guide the analyses on the state of play in the three subfields we selected in Part II – renewable energy (Chapter 4), fossil fuel subsidy reform (Chapter 5), and carbon pricing (Chapter 6).

Juxtaposing the themes of coherence and management leads us to six generic ways of characterizing the institutional complex of a subfield of the climate-energy nexus (Table 2.1). 'Generic' also implies that the intermediate concepts (coexistence or duplication, coordination) describe a middle ground that serves as a proxy for an actual continuum between synergy and conflict.

Table 2.1 *Dimensions for assessing the institutional complex of a subfield of the climate-energy nexus.*

Coherence Management	High	Medium	Low
Non-managed relationship	Synergy	Coexistence / Duplication	Conflict
Managed relationship	Division of labour	Coordination	Competition

The terms in the upper row of Table 2.1 are straightforward and intuitive and reflect when we can speak of high or low coherence, in terms of the three core dimensions we introduced earlier. Coherence exists if institutions converge on a core norm and exert governance functions accordingly and synergistically, while covering a large set of complementary audiences without too many gaps or duplications. Conflict, on the other side of the spectrum, refers to a strong divergence on the core norm and respective contradictions, rivalries, and gaps in the exercise of governance functions for various audiences.

The three types of managed relationships (division of labour, coordination, competition) are understood correspondingly. We define an institutional division of labour as the deliberate and continuous sharing of governance functions and norms among institutions for complimentary memberships. This definition chimes with Young (1996) and his understanding of 'embedded institutions' that work on different *issues*, but share overarching principles and practices, and 'nested institutions' that work on the same issue, but differ in *functional* scope or *geographical* domain and are 'folded into broader institutional frameworks'.

Coordination, as a middle ground, would mark deliberate management attempts in a policy subfield that rather facilitate an ad-hoc or case-based sharing of specific governance functions.

Competition, finally, describes deliberate attempts to manage institutional linkages that are either not successful or that are only meant to facilitate a legal or political frame in which institutions vie for the same governance task and / or audience. 'Vying' here relates to competition for predominance, either in a specific sector or geographical level or across sectors or levels.

2.3.3 Legitimacy

While the previous two themes, coherence and management, focused on the shape of complexity among institutions, this and the next section concentrate on the consequences of this complexity for the affected institutions, their operation, and their goal-fulfilment. We focus on two distinct yet related concepts: legitimacy and effectiveness. This is because, on a fundamental level, institutions, and institutional complexes or architectures as a whole, need to be both legitimate and effective to function well in the long term (Andresen and Hey 2005). In the following we argue why we need new approaches to studying legitimacy and effectiveness that take into account the wider context that institutions find themselves in. Specifically, we show how both legitimacy and effectiveness are multidimensional concepts that need to be investigated through multiple methods.

The evaluative theme of legitimacy has been discussed extensively in the literatures on global governance and institutional analysis (Suchman 1995; Hurd 1999; Bernstein 2005; Buchanan and Keohane 2006; Clark 2007; Keohane 2011). Broadly it refers to 'the acceptance and justification of shared rule by a community' (Bernstein 2005, 142). A legitimate institution thus rules with authority as it has obtained support for its operation. Legitimacy is important for global governance, as the alternative tools for generating compliance with shared rules in the international system – inducement or coercion – 'are often unavailable, in short supply, or costly to use' (Bernstein 2011, 20). Global governance institutions that seek to exercise legitimate power must thus gain acceptability and credibility amongst the communities that they seek to govern.

Legitimacy can be studied through different approaches. Two common approaches are normative and sociological legitimacy. While the former examines whether the authority in question performs according to some pre-defined standards, the latter is concerned with whether the authority is perceived as having the right to rule amongst those it seeks to govern (Buchanan and Keohane 2006). Given these different approaches, scholarly evaluations of legitimacy may diverge depending on whether the focus of the study is to examine justifications of operations or the acceptability of those claims to a given constituency.

In sum, legitimacy is a multidimensional concept that can be evaluated in different ways. Institutional complexity adds another challenge to this evaluation due to the interplay between different sets of institutions – each with their own audiences, objectives, processes, and consequences. In particular, institutional complexity blurs the question of attribution due to the interdependence between institutions. With some institutions being placed in a subfield crowded with competing institutions with overlapping mandates (Abbott et al. 2016), it is very difficult, if not outright impossible, to disentangle the legitimacy of individual institutions. Conventional methods of assessing legitimacy only by looking at institutions individually against a set of criteria thus provides an insufficient perspective for studying both legitimacy and complexity (Alter and Meunier 2009).

With these caveats in mind, we choose a different and novel approach to assess the legitimacy of the climate-energy nexus. We argue that it is necessary to draw on a mix of the normative and sociological approaches in order to understand legitimacy under institutional complexity. Specifically, we highlight an aspect of legitimacy that has to date received little attention in the literature – namely how key stakeholders value this concept. To study this phenomenon, we start from the normative literature but then rely on interviews and questionnaires to explore how stakeholders assess legitimacy.

Our focus on stakeholder perceptions is grounded in two considerations. The first of these arises from an academic debate on how institutional complexity affects actors' choices in international cooperation in general, and political decision-making in particular (Alter and Meunier 2009). The second one stresses the policy implications of institutional complexity, concretely the notion of an era in which global governance institutions compete over members and resources (Andresen and Hey 2005; Karlsson-Vinkhuyzen and McGee 2013). While actors join institutions based on strategic considerations, global governance institutions need to be legitimate and must gain acceptability and credibility amongst the communities they seek to govern to achieve their objectives. It is therefore important to empirically study the views of those actors that global governance institutions seek to govern. In Part III we therefore examine if and how legitimacy assessments vary between key audiences (c.f. Nasiritousi et al. 2016).

Who these legitimacy-granting actors are has been the subject of some debate in the literature. While traditionally only states, as the central international actors, were seen as members of the community that grant legitimacy to international institutions, the boundaries of this community have shifted as nonstate actors have risen in prominence in recent years. Thus, international institutions now need to be perceived as legitimate not only by states, but also by 'democratic publics', including a range of stakeholders (Symons 2011, 2561; Gronau and Schmidtke 2016). This directly connects to the core dimension of membership that we introduced earlier for assessing coherence. Institutional membership may not necessarily be identical with a legitimacy audience, but it is an important and defining dimension for identifying such an audience.

Based on these considerations, which dimensions are suitable to examine legitimacy under conditions of institutional complexity? Or, put in perceptional terms, which dimensions help us examine how institutional complexity affects stakeholders' assessments of key institutions of the climate-energy nexus? Chapter 7 will develop such a set of dimensions in further detail and apply them to the renewable energy subfield of the climate and energy nexus. In the following we will only briefly sketch how, and based on which theoretical assumptions, we derive these dimensions.

While the general concept of legitimacy is contested, the literature is in relative agreement about the institutional qualities that contribute to legitimacy

Box 2.7
Dimensions of Legitimacy

- Expertise;
- Inclusion of all appropriate actors;
- Procedural (decision-making) fairness;
- Transparency;
- Accountability;
- Output (what is produced);
- Outcome (the effect the output has on its members);
- Impact (the effect the output has on problem-solving);
- Distributive fairness (distributing benefits to members fairly).

(Karlsson-Vinkhuyzen and Vihma 2009; Mena and Palazzo 2012; Scholte and Tallberg 2018). Yet, the extent to which stakeholders believe that different institutions are fulfilling these criteria has thus far not been empirically explored. Therefore, based on the literature on what makes an institution legitimate, we derive a set of dimensions for assessing the normative legitimacy of international institutions. These include both process-related (input) dimensions as well as outcome-related (output) ones (see Box 2.7, and Chapter 7 for a detailed introduction of each dimension):

When applying these dimensions in Chapter 7, we map how different audiences assess these nine dimensions of legitimacy that we derived from the normative literature for five key institutions in the renewable energy subfield. This approach hence occupies a middle ground between the normative approach (whereby the researcher assesses an institution based on normative criteria) and the sociological approach, which focuses on the legitimacy perceptions of audiences (cf. Agné 2018). Gaining an understanding of these perceptions is important for institutions that seek to tailor their legitimation strategies toward different audiences (Bäckstrand and Söderbaum 2018).

There are two main explanations for why legitimacy perceptions may diverge between different types of stakeholders: according to the traditional view of legitimacy in IR, the congruence model, actors make informed assessments about institutions. In this understanding, differences in legitimacy perceptions stem from variations in legitimacy demands between groups of stakeholders. According to the cognitive model, in contrast, actors rely on mental shortcuts to make assessments. This view builds on insights from the field of cognitive psychology, and highlights that legitimacy perceptions are rooted in cognitive schemata and heuristics. This means that differences in legitimacy perceptions instead stem from variations in the heuristics that different actors use to make the assessment (Lenz and Viola 2017).

The latter model is expected to be of use particularly under conditions of institutional complexity since actors (even if they are experts) cannot be expected to make a fully informed assessment as implied by the congruence model. This is in line with research that has found that 'complexity forces bounded rationality on actors' (Alter and Meunier 2009, 17), and that actors rely on heuristics in the face of overwhelming information. Chapter 7 does not seek to explain how legitimacy assessments are made but instead assumes that differences in perceptions can stem both from differences in norms and values, as well as differences in experiences and thereby heuristics used to form opinions. The chapter thus maps the differences in assessments between key institutions and stakeholder groups to provide novel empirical insights into the normative legitimacy of these institutions.

2.3.4 Effectiveness

As with legitimacy, the evaluative theme of effectiveness is multidimensional. Generally speaking, effectiveness can be conceptualized as the level of goal attainment by an institution (Bernauer 1995). In the literatures on International Relations and Comparative Politics, effectiveness is typically operationalized through three dimensions, as the output, outcome, or impact of institutions (Underdal 2002; Tallberg et al. 2016; see Box 2.8). Whereas output is a process-based dimension that looks at the narrow governance functions of an institution, and outcome examines the political impacts or behavioural change resulting from the operations of the institution, impact looks at issues of problem resolution and the extent to which the institution has contributed to welfare enhancement (Gutner and Thompson 2010).

These three dimensions of effectiveness correspond to three of the nine dimensions of legitimacy we just introduced. This overlap suggests that, while legitimacy and effectiveness are distinct concepts, they are also strongly interrelated – e.g. with (perceived) high levels of effectiveness positively impacting the legitimacy (perceptions) of the institution in question.

Since output focuses on the immediate performance of an institution, it is the most observable of the three dimensions and does not have to consider issues of causation. This may make it the preferred operationalization of effectiveness from

Box 2.8
Dimensions of Effectiveness

- Output (what is produced);
- Outcome (the effect the output has on its members);
- Impact (the effect the output has on problem-solving).

a methodological viewpoint, but it only captures potential effectiveness, as it does not consider the actual consequences of the output. Impact, conversely, is difficult to measure, but captures how well the institution changes target indicators – and is hence in a better position to answer questions concerning whether and how the institution is able to solve the problems that it has been designed to tackle (Underdal 2002). Ideally, an investigation into all three dimensions of effectiveness provides a fuller picture of the performance of the institution. An analysis of how outputs link to outcome and impact would provide valuable insights into the actual effectiveness of the institution, especially if such an analysis also takes into account unintended side effects of the institution's work.

Another complicating factor in assessing effectiveness is that international institutions typically have several, and sometimes conflicting, goals or core norms (for example, there exist tensions between alleviating energy poverty and promoting sustainable energy). Therefore, similar to sociological legitimacy where legitimacy demands may vary between actors, evaluations of effectiveness may vary amongst constituencies depending on which goal-fulfilment is favoured. This is what Gutner and Thompson (2010) refer to as the 'eye of the beholder' problem, as effectiveness may be assessed differently by members of an institution and outside stakeholders, but sometimes also between members if interests diverge.

Chapter 8 turns to the question of effectiveness to evaluate the three subfields of the climate-energy nexus. Here we rely on a mixed approach of assessing three dimensions of effectiveness – output, outcome, and impact – of the institutions at the meso level by combining interview data with experts in the field and the researchers' own assessments of how outputs relate to outcomes and impacts. The chapter discusses these three dimensions and highlights the difficulties involved in determining the outcomes and impacts of institutions at the meso level given the range of confounding factors and the problem of attribution under institutional complexity.

Nevertheless, the chapter offers a transparent evaluation of the effectiveness of the subdomains – by discussing the outputs at the meso level, in terms of, for example, data and reports, capacity building, standard-setting financing, and implementation of projects, and then discussing how these may (or may not) be linked to observed outcomes and impacts. The value of the analysis lies in the examination of how institutional complexity affects the effectiveness of the subfields and the identification of management options that may be useful in addressing some of the existing bottlenecks.

2.4 Conclusions

The previous pages introduced our understanding of major terms (nexus, institutions, organizations, regimes, initiatives, networks) and analytical levels (macro, meso, micro) as well as a novel combination of evaluative themes and their dimensions for a thorough and comparative analysis of the macro, meso, and micro levels of the energy nexus:

- Coherence – in terms of convergence on a core norm; and balanced distribution and synergy on governance functions, and membership as well as underlying causal mechanisms;
- Management – in terms of agents, levels, and consequences for normative, functional, or membership-related convergence;
- Legitimacy – in terms of perception of institutional audiences of process- and outcome-based criteria;
- Effectiveness – in terms of the output, outcome, and impact-related consequences of institutional complexity at micro and meso levels.

These four themes are not meant to present a silver bullet toward studying a governance nexus or institutional complexity in general. First, they are not mutually exclusive. For instance, management is defined as deliberate attempts to provide more coherence, and effectiveness is framed as an important cornerstone of legitimacy. Second, they are not exhaustive, since there are a series of other dimensions which could also be assessed in such a nexus, such as underlying discourse coalitions or social networks.

These qualifications notwithstanding, the four evaluative themes, their dimensions and criteria, provide an detailed framework for an in-depth analysis of a governance nexus. They help us examine a variety of important questions in a comparative research design, combining a high level of ambition with feasibility and novelty.

The following chapters will illustrate these qualities when applying the analytical framework to our three case studies. Chapters 3–6 will provide an encompassing application of the framework's criteria for coherence and management. Chapter 3 offers overarching perspectives on macro-level coherence for the nexus as a whole and for the meso-level coherence for the three subfields. Chapters 4–6 then delve into each subfield and provide further details on meso-level coherence as well as analyses of micro-level coherence and management attempts. Chapters 7 and 8 follow with in-depth applications of the legitimacy and effectiveness criteria of the framework before our conclusions in Chapter 9 tie together our various results in a comparative and comprehensive manner, summarizing the strength of our multi-dimensional approach.

2.5 References

Abbott, K. W. 2012. The Transnational Regime Complex for Climate Change. *Environment & Planning C: Government & Policy* 30(4), 571–590.

Abbott, K. W. and Snidal, D. 2009a. Strengthening International Regulation through Transnational New Governance: Overcoming the Orchestration Deficit. *Vanderbilt Journal of Transnational Law* 42(2), 501–578.

Abbott, K. W. and Snidal, D. 2009b. The Governance Triangle: Regulatory Standards Institutions and the Shadow of the State. In *The Politics of Global Regulation*, edited by W. Mattli and N. Woods. Princeton: Princeton University Press.

Abbott, K. W., Green, J. F., and Keohane, R. O. 2016. Organizational Ecology and Institutional Change in Global Governance. *International Organization* 70(2), 247–277.

Agné, H. 2018. Legitimacy in Global Governance Research: How Normative or Sociological Should It Be? In *Legitimacy in Global Governance: Sources, Processes, and Consequences*, edited by J. Tallberg, K. Bäckstrand, and J. A. Scholte. Oxford: Oxford University Press.

Alter, K. J. and Meunier, S. 2009. The Politics of International Regime Complexity. *Perspectives on Politics* 7(1), 13–24.

Andresen, S. and Hey, E. 2005. The Effectiveness and Legitimacy of International Environmental Institutions. *International Environmental Agreements: Politics, Law and Economics* 5(3), 211–226.

Bäckstrand, K. and Söderbaum, F. 2018. Legitimation and Delegitimation in Global Governance: Discursive, Institutional, and Behavioural Practices. In *Legitimacy in Global Governance: Sources, Processes, and Consequences*, edited by, J. Tallberg, K. Bäckstrand, and J. A. Scholte. Oxford: Oxford University Press.

Bernauer, T. 1995. The Effect of International Environmental Institutions: How We Might Learn More, *International Organization* 49(2), 351–377.

Bernstein, S. 2005. Legitimacy in Global Environmental Governance. *Journal of International Law and International Relations* 1(1–2), 139–166.

Bernstein, S. 2011. Legitimacy in Intergovernmental and Non-State Global Governance. *Review of International Political Economy* 18(1), 17–51.

Biermann, F. and Kim, R. E. Forthcoming. Introduction. In *Global Governance for the Earth: Transforming Institutional Architectures in the Anthropocene*, edited by F. Biermann and R. E. Kim. Cambridge, MA: The MIT Press.

Braithwaite, J. and Drahos, P. 2000. *Global Business Regulation*. Cambridge: Cambridge University Press.

Buchanan, A. and Keohane, R. O. 2006. The Legitimacy of Global Governance Institutions. *Ethics & International Affairs* 20(4), 405–437.

Clark, I. 2007. *Legitimacy in International Society*. Oxford: Oxford University Press.

Conca, K. 2006. *Governing Water: Contentious Transnational Politics and Global Institution Building*. Cambridge: MIT Press.

Eberlein, B. et al. 2014. Transnational Business Governance Interactions: Conceptualization and Framework for Analysis. *Regulation & Governance* 8(1), 1–21.

Finnemore, M. 1996. *National Interests in International Society*. Ithaca, NY: Cornell University Press.

Gehring, F. and Faude, B. 2014. A Theory of Emerging Order within Institutional Complexes. *Review of International Organizations* 9(4), 471–498.

Gehring, T. and Oberthür, S. 2006. Comparative Empirical Analysis and Ideal Types of Institutional Interaction. In *Institutional Interaction in Global Environmental Governance: Synergy and Conflict among International and EU Policies*, edited by S. Oberthür and T. Gehring. Cambridge, MA: MIT Press.

Gehring, T. and Oberthür, S. 2009. The Causal Mechanisms of Interaction between International Institutions. *European Journal of International Relations* 15(1), 125–156.

Gronau, J. and Schmidtke, H. 2016. The Quest for Legitimacy in World Politics – International Institutions' Legitimation Strategies. *Review of International Studies* 42(3), 535–557.

Gutner, T. and Thompson, A. 2010. The Politics of IO Performance: A Framework. *The Review of International Organizations* 5(3), 227–248.

Helfer, L. 2009. Regime Shifting in the International Intellectual Property System. *Perspectives on Politics* 7(1), 39–44.

Hurd, I. 1999. Legitimacy and Authority in International Politics. *International Organization* 53(2), 379–408.

Karlsson-Vinkhuyzen, S. I. and McGee, J. 2013. Legitimacy in an Era of Fragmentation: The Case of Global Climate Governance. *Global Environmental Politics* 13(3), 56–78.

Karlsson-Vinkhuyzen, S. I. and Vihma, A. 2009. Comparing the Legitimacy and Effectiveness of Global Hard and Soft Law: An Analytical Framework. *Regulation and Governance* 3(4), 400–420.

Keohane, R. O. 1989. *International Institutions and State Power. Essays in International Relations Theory*. Boulder, CO: Westview Press.

Keohane, R. O. 1993. The Analysis of International Regimes. Towards a European-American Research Programme. In *Regime Theory and International Relations*, edited by V. Rittberger. Oxford: Clarendon Press.

Keohane, R. O. 2011. Global Governance and Legitimacy. *Review of International Political Economy* 18(1), 99–109.

Lenz, T. and Viola, L. A. 2017. Legitimacy and Institutional Change in International Organisations: A Cognitive Approach. *Review of International Studies* 43(5), 939–961.

Mena, S. and Palazzo, G. 2012. Input and Output Legitimacy of Multi-Stakeholder Initiatives. *Business Ethics Quarterly* 22(3), 527–556.

Nasiritousi, N., Hjerpe, M., and Bäckstrand, K. 2016. Normative Arguments for Non-State Actor Participation in International Policymaking Processes: Functionalism, Neocorporatism or Democratic Pluralism? *European Journal of International Relations* 22(4), 920–943.

Oberthür, S. 2009. Interplay Management: Enhancing Environmental Policy Integration among International Institutions. *International Environmental Agreements: Politics, Law and Economics* 9(4), 371–391.

Oberthür, S. and Gehring, T. 2006. Institutional Interaction in Global Environmental Governance: The Case of the Cartagena Protocol and the World Trade Organization. *Global Environmental Politics* 6(2), 1–31.

Oberthür, S. and Stokke, O. S. 2011. *Managing Institutional Complexity: Regime Interplay and Global Environmental Change*. Cambridge, MA: The MIT Press.

Ruggie, J. G. 1998. Introduction: What Makes the World Hang Together? Neo-Utilitarianism and the Social Constructivist Challenge. In *Constructing the World Polity: Essays on International Relations*, edited by J. G. Ruggie. London: Routledge.

Schneider, C. J. and Urpelainen, J. 2013. Distributional Conflict Between Powerful States and International Treaty Ratification. *International Studies Quarterly* 57:1, 13–27.

Scholte, J. A. and Tallberg, J. 2018. Theorizing the Institutional Sources of Global Governance Legitimacy. In *Legitimacy in Global Governance: Sources, Processes and Consequences*, edited by J. Tallberg, K. Bäckstrand, and J. A. Scholte. Oxford: Oxford University Press.

Stokke, O. S. 2001. *The Interplay of International Regimes: Putting Effectiveness Theory to Work? FNI Report 10/2001*. Lysaker: Fridtjof Nansen Institute.

Suchman, M. 1995. Managing Legitimacy: Strategic and Institutional Approaches. *The Academy of Management Review* 20(3), 571–610.

Symons, J. 2011. The Legitimation of International Organisations: Examining the Identity of the Communities that Grant Legitimacy. *Review of International Studies* 37(5), 2557–2583.

Tallberg J., Sommerer, T., Squatrito. T., and Lundgren, M. 2016. The Performance of International Organizations: An Output-Based Approach. *Journal of European Public Policy* 23(7), 1077–1096.

Tosun, J., Koos, S., and Shore, J. 2016. Co-Governing Common Goods: Interaction Patterns of Private and Public Actors. *Policy and Society* 35(1), 1–12.

Underdal, A. 2002. One Question, Two Answers. In *Environmental Regime Effectiveness: Confronting Theory with Evidence*, edited by E. Miles et al. Cambridge, MA: MIT Press.

Van de Graaf, T. 2013. Fragmentation in Global Energy Governance: Explaining the Creation of IRENA. *Global Environmental Politics* 13(3), 14–33.

Van de Graaf, T. and Zelli, F. 2016. Actors, Institutions and Frames in Global Energy Politics. In *The Palgrave Handbook of the International Political Economy of Energy*, edited by T. Van de Graaf, B. K. Sovacool, F. Kern, A. Ghosh, and M. T. Klare. London: Palgrave Macmillan.

Young, O. R. 1996. Institutional Linkages in International Society: Polar Perspectives. *Global Governance* 2(1), 1–24.

Zelli, F. 2010. *Conflicts among International Regimes on Environmental Issues*. A Theory-driven Analysis. PhD Dissertation. Tübingen: Eberhard-Karls University.

Zelli, F. 2018. Global Governance. In *Concise Oxford Dictionary of Politics and International Relations*. 4th edition, edited by G. W. Brown, I. McLean, and A. McMillan. Oxford: Oxford University Press.

Zelli, F., Gupta, A., and van Asselt, H. 2013. Institutional Interactions at the Crossroads of Trade and Environment: The Dominance of Liberal Environmentalism? *Global Governance* 19(1), 105–118.

3

Mapping the Institutional Complex of the Climate-Energy Nexus

LISA SANDERINK, PHILIPP PATTBERG, AND OSCAR WIDERBERG

3.1 Introduction

Global energy challenges and responses to climate change are intrinsically intertwined. Efforts to achieve the Sustainable Development Goal on energy (SDG 7) – to provide '*access to affordable, reliable, sustainable and modern energy for all*' (United Nations 2015) – will affect the possibility to reach the goals set out in the Paris Agreement under the United Nations Framework Convention on Climate Change (UNFCCC) to keep global warming under 2 degrees, and vice versa. Such efforts are thus interdependent and situated in a 'climate-energy nexus'.

In this chapter, we map the increasingly dense institutional complex of governance institutions occupying the climate-energy nexus. In global climate governance, international and transnational institutions have proliferated since the adoption of the UNFCCC in 1992, which has led to an increasingly fragmented global climate governance architecture (Bäckstrand 2008; Biermann et al. 2009; Keohane and Victor 2011; Abbott 2012; Bulkeley et al. 2014; Falkner 2014). Similarly, global energy governance has been characterized as fragmented (e.g. Dubash and Florini 2011; Van de Graaf 2013; Escribano 2015), partly due to the diversity of governance efforts involved and the way it deals with different energy sources (e.g. coal, gas, solar, and wind), and challenges (energy security, energy access and environmental sustainability).

Despite the interdependence between climate change and energy governance, scholars studying institutional structures often only focus on one issue area. For example, in an effort to understand institutional complexity of global climate change governance, and the causes and consequences thereof, several studies introduced mappings of the broader institutional complex. Keohane and Victor (2011) discussed the regime complex for climate change by demonstrating a plethora of international state-based governance arrangements, and evaluated the emerging regime complex as 'loosely coupled', with institutions that are not integrated or arranged in a clear hierarchy (Keohane and Victor 2011, 9). Others have mapped climate governance institutions beyond the international realm, focusing on transnational and private climate governance (e.g. Pattberg and Stripple 2008; Abbott 2012; Bulkeley et al. 2014; Hale and Roger 2014). Subsequent research sought to combine both spheres (Widerberg et al. 2016), operationalizing a heuristic framework that had been developed by Abbott and Snidal (2009a; 2009b; Abbott 2012).

Similarly, a growing body of literature has mapped the global energy complex (e.g. Suding and Lempp 2007; Kerebel and Keppler 2009; Lesage et al. 2010; Colgan et al. 2012; Sovacool and Florini 2012; Leal-Arcas and Filis 2013; Wilson 2015; Escribano 2015). However, these mappings led to strikingly different results, with the number of governance efforts ranging from six, identified by Kerebel and Keppler (2009), to fifty, identified by Sovacool and Florini (2012) (Van de Graaf and Colgan 2016). Like the institutional complex itself, the mapping efforts have been rather fragmented in terms of focus on energy source, or type of institution. Whereas a number of studies exclusively focus on oil and gas (e.g. Kerebel and Keppler 2009), others target renewable energy sources (e.g. Barnsley and Ahn 2014). Additionally, while some mappings are restricted to intergovernmental organizations (e.g. Wilson 2015), others also include nongovernmental organizations; and hybrid or public–private institutions (Sovacool and Florini 2012). Sanderink et al. (2018), finally, merged these different criteria and introduced a novel and comprehensive mapping of global energy governance, following a methodology similar to Widerberg et al. (2016).

What is missing to date is an integrated mapping and coherent analysis of the institutional complex governing the nexus between climate change and energy. Consequently, this chapter identifies the institutions that address both challenges simultaneously. The aims of this chapter are twofold: first, we provide a pioneering mapping and analysis of the climate-energy institutional nexus. Second, we introduce novel empirical data and input for the three case studies on the subfields of renewable energy, fossil fuel subsidy reform, and carbon pricing (see Chapters 4–6).

Concretely, this chapter analyzes the climate-energy nexus along the distinction between the macro level (the overall nexus) and the meso level (the various subfields within the broader climate and energy institutional complex). Each level is scrutinized along a number of analytical questions, including the major dimensions that were introduced in Chapter 2. When were the institutions established, and how did the institutional complex develop over time? What types of institutions populate the climate-energy nexus? Who are the institutional members to the institutions, and how are these connected? What kind of governance functions do these institutions fulfil? What is their thematic focus?

The chapter proceeds as follows: Section 3.2 introduces the methodology and data collection approach. Section 3.3 provides an introduction to the three subfields; Sections 3.4 and 3.5 then provide the mapping and analysis of the macro and meso levels of the climate-energy nexus. Section 3.6 concludes with a set of final remarks and sets the stage for the subsequent chapters.

3.2 Methodology

Mapping the institutional complex that governs the climate-energy nexus and the three subfields is carried out in two steps. First, compiling a database that includes the institutions actively addressing the climate-energy nexus; and second, visualizing and analyzing the data. The next subsections describe these two steps in more detail, including our dataselection criteria and data analysis.

3.2.1 Data Selection

The database consists of institutions that govern both climate change and energy. In line with our definition of global governance (see Chapter 2), our criteria for inclusion and exclusion are based on previous work by the CONNECT project,[1] which includes institutions that are (i) intergovernmental or transnational, which not only have the (ii) intentionality to steer policy and the behaviour of their members or a broader community, but also explicitly refer to a (iii) common governance goal, to be accomplished by (iv) significant governance functions (Widerberg et al. 2016). For the climate-energy nexus as a whole, the overarching governance goals are twofold. On the one hand, institutions strive to mitigate climate change, i.e. to stabilize greenhouse gas concentrations in the atmosphere at a level that will prevent dangerous human interference with the climate system. On the other hand, institutions in the nexus adhere to the proposition that decarbonizing the energy sector is essential to combat climate change. Taken together, the governance goal that unites institutions in the nexus is greenhouse gas mitigation through a transformation toward low-carbon or fossil-free energy systems, excluding those focusing on carbon capture and storage (CCS) and nuclear power.

For populating the database, we used two key sources. First, the CONNECT project's database on global climate change governance (Widerberg et al. 2016), and secondly, the CONNECT dataset on global energy governance (Sanderink

[1] The CONNECT project (1) takes stock of the existing level of fragmentation across a number of issue areas in global environmental governance, (2) explains the causes of fragmentation of global governance architectures, (3) analyzes the implications of fragmentation across different scales of governance, and finally (4) suggests policy responses to increased fragmentation. The project was funded with an NWO Innovational Research Incentives Schemes Vidi Grant and is hosted by the Environmental Policy Analysis Department at the Institute for Environmental Studies, Vrije Universiteit Amsterdam.

et al. 2018). The database on climate governance builds on previous case studies (particularly Hoffmann 2011; Bulkeley et al. 2014; Hale and Roger 2014) and a thorough assessment of two online databases: the Climate Initiative Platform (climateinitiativesplatform.org) and the Non-State Actor Zone for Climate Action (NAZCA, climateaction.unfccc.int). Similarly, the dataset on global energy governance includes previous mapping exercises (Suding and Lempp 2007; Kerebel and Keppler 2009; Lesage, Van de Graaf, and Westphal 2010; Colgan, Keohane, and Van de Graaf 2012; Sovacool and Florini 2012; Escribano 2015; Wilson 2015), complemented with data from the Climate Initiatives Platform, NAZCA, and the Portal on Cooperative Initiatives (unfccc.int/focus/mitigation/items/7785.php). From these two key sources, we selected those institutions focusing on climate and energy simultaneously for the climate-energy nexus database. Finally, a small number of institutions were added based on complementary discussions with experts on climate-energy overlaps.[2]

3.2.2 Data Analysis

The data we retrieved for the selected institutions includes: name of institution, starting year, membership data, governance functions, and thematic focus. Based thereon, the mappings of the climate-energy nexus and the subfields can be illustrated in various ways. First, we started with a timeline demonstrating how, when, and in which context the institutional complex governing the climate-energy nexus emerged.

Second, for an overview of the institutional complex and the subfields, we used a heuristic framework developed by Abbott and Snidal (2009a; 2009b; Abbott 2012) for mapping global governance architectures, called 'governance triangle'. We situated the institutions from our database in the governance triangle according to their membership, i.e. the type of their constituent actors: public, firm, and/or civil society organization (CSO). The public category includes individual states (or their governmental agencies, respectively), groups of states, international organizations (IOs), cities, or regions. The firm category comprises (groups of) firms, industry associations, and investors. Finally, CSOs include nongovernmental organizations (NGOs), other organizations that represent civil society, and networks and coalitions of CSOs. Based on the three actor categories, the governance triangle is divided into seven zones. Institutions in zones 1–3, the vertex zones, are constituted by a single type of actor. Those in zones 4–6, the quadrilateral zones,

[2] Complementary discussions took place at CLIMENGO project meetings. CLIMENGO is a research project that aims to map the institutional complexity of global climate and energy governance, evaluate its effectiveness and legitimacy, and develop a knowledge base for decision makers (www.climengo.eu). The experts include project members Karin Bäckstrand, Jakob Skovgaard, Harro van Asselt, and Fariborz Zelli.

include two types of actors. Finally, the institutions in the central zone 7 are those that involve all three types, also called multi-stakeholder institutions. In the respective triangles that follow in this chapter, we will also use greyscale to mark these different actor patterns.

Third, for each institution in the dataset we collected membership data using a methodology developed by the CONNECT project (Widerberg et al. 2016). Members are defined as '*actors with the formal position to influence the rules, norms, operations or performance of an institution*' (Widerberg et al. 2016, 19). Moreover, they may gain benefits from their membership through access to their institution's network, and due to material, reputational, or other types of benefits that the institution is expected to yield. On the other hand, we excluded 'supporting' organizations or countries from the dataset. These could be organizations that merely support an institution's values, rules, norms, or mission by way of a public statement or endorsement. Moreover, we only accounted for collective actors such as countries, companies, cities, regions, or NGOs as members, i.e. not their individual representatives.

To tackle further ambiguities, we used four rules when collecting the membership data. First, for institutions engaging in pledges and commitments, only the organizing or lead institutions have been included (e.g. DivestInvest). Second, for institutions engaged in certifications and registries, we only considered those organizations with the power to hand out or change the certificates as well as those collecting the data for the registries (e.g. Gold Standard). Third, in cases where we have not been able to establish authority due to a networked mode of governance (e.g. in city networks) we treated all participants as members (e.g. Covenant of Mayors). Fourth, for institutions where a member may join a decision-making body, such as the steering committee or board of directors, all members with such privileges have been included (e.g. International Emissions Trading Association).

The membership data we gathered, based on these rules and criteria, enabled us to explore in more detail who the key actors in the climate-energy nexus are. We summarized these in a network diagram (see Section 3.4.2.2), with nodes representing institutions and members and edges indicating which members are shared among the institutions. This type of visualization highlights which institutions are central and which countries are best connected in terms of membership.

Fourth, another form of visualization, the governance decagon, displays the governance functions that individual institutions perform. The decagon is divided into ten segments representing four different governance functions, which were also introduced in Chapter 2, and combinations of these. The governance function 'standards and commitments' refers to rule-making and implementation schemes, involving mandatory compliance, standards for measurements and disclosure, and voluntary and private standards and commitments. The function 'operational activities' comprises, for example, technology research and development, pilot projects, demonstration and deployment activities, skills enhancement, and best practices. The 'information and networking' governance function encompasses information-sharing forums and networking, such as technical consulting, training, and information services to build capacity, share knowledge, and support local governments. Finally, the governance function 'financing' refers to any operational activities that involve financing schemes. A number of institutions employ more than two governance functions, but for the purpose of clarity the decagon is restricted to two functions per institution. In such cases, the authors had to make a final call on how the institutions should be classified, often based on a judgement of which governance functions appear dominant, while additional governance functions may be elaborated on in the analyses of Chapters 4–6.

Finally, after compiling and evaluating the dataset, we determined for each institution individually which thematic focus is most relevant. After studying the institutions' websites, we distinguished eight foci: (1) increasing the uptake of renewable energy; (2) pricing and trading of carbon emissions; (3) reforming harmful fossil fuel subsidies; (3) improving energy efficiency; (4) expanding worldwide access to low-carbon energy (services); (5) financing climate mitigation actions and decarbonizing investments; (7) developing low or zero carbon technologies; and (8) transitioning toward clean fuels in the transport sector. The institutions that govern toward these thematic foci form the three subfields that we scrutinize in this volume (renewable energy, fossil fuel subsidy reform, and carbon pricing), and five additional ones in the climate-energy nexus. These subfields should not be considered as silos: first, since institutions can address multiple thematic foci, and second, these thematic foci are crosscutting. For example, clean technologies can include fuel-efficient engines for the transport sector, and financing mechanisms may be designed to make clean cooking appliances accessible. As a consequence, subfields in the climate-energy nexus can show overlaps in terms of institutions.

3.3 Three Subfields of the Climate-Energy Nexus: Renewable Energy, Fossil Fuel Subsidy Reform, and Carbon Pricing

Various themes and activities fall into the intersection of the climate and energy domains, for instance promoting energy efficiency and respective technologies, contributing to worldwide access to renewable and low-carbon energy, and introducing non-fossil fuels in the transport sector. The institutions centred around these and other thematic foci constitute various subfields within the institutional complex for the climate-energy nexus. The three subsequent chapters of this volume put particular emphasis on three of these subfields and analyze them at the meso and micro levels: renewable energy, fossil fuel subsidy reform, and carbon pricing. In what follows, we briefly introduce these three subfields and outline some of the key institutions therein.

The subfield focusing on renewable energy comprises institutions that support uptake, installation, technologies, and information-sharing on renewable energy (see Sanderink, Chapter 4). Renewables play a significant role in the world's trajectory to sustainable development. An enhanced uptake helps to alleviate the increasing scarcity of energy sources and reduce air pollution and greenhouse gas emissions. The growing renewable energy sector is also highly compatible with decentralized and small-scale deployment efforts to expand energy access. Yet, despite environmental concerns raised in the early 1990s (e.g. UNCED 1992), it was not until the turn of the millennium that renewable energy started to receive increased attention (Röhrkasten 2015). UN work on renewable energy remains weakly developed, and, arguably as a result, a series of institutions on this topic have emerged outside the UN framework. Important intergovernmental institutions are the International Renewable Energy Agency (IRENA) and the International Energy Agency (IEA). Additionally, a range of multi-stakeholder partnerships were established, such as the Renewable Energy and Energy Efficiency Partnership (REEEP) and the Renewable Energy Policy Network for the 21st Century (REN21). Moreover, minilateral institutions such as the G8 and G20 are seen as key in promoting renewables (e.g. Florini and Sovacool 2009; Colgan and Van de Graaf 2014).

The second subfield under scrutiny in this volume consists of institutions that support the reform of harmful fossil fuel subsidies. Fossil fuels are still heavily subsidized around the globe and therefore remain highly competitive, with respective infrastructures kept in place. Yet, the urgency of reforms has been increasingly recognized (see Verkuijl and van Asselt, Chapter 5). Research shows that removing fossil fuel subsidies connected to consumption in twenty-seven countries between 2013 and 2020 would lead to an 8 per cent reduction in global greenhouse gas emissions (Burniaux and Chateau 2014). International cooperation efforts play an important role in worldwide fossil fuel subsidy reform, and can be traced back to the G20 summit in Pittsburgh in 2009. The meeting led to a first international commitment to address fossil fuel subsidies (G20 2009), closely followed by a similar pledge by twenty-one members of the Asia Pacific Economic Cooperation (APEC) (APEC 2009). Subsequently, several additional institutions became active in the field such as the IEA, Friends of Fossil Fuel Subsidy Reform (Friends), the International Monetary Fund (IMF), and the Global Subsidies Initiative (GSI).

Finally, the third subfield consists of institutions that aim at putting a price on carbon, facilitating a trade system, and providing a system for offsetting emissions.

Putting a price on carbon is seen as a fundamental solution to climate change (e.g. Sterner and Coria 2011; Tol 2011). This argument rests upon the idea that climate change is best addressed by creating an incentive for individuals to reduce emissions with the help of a price signal (see Skovgaard and Canavan, Chapter 6). Carbon taxes, emission trading systems, and mechanisms to offset emissions are well-known examples of these market-based instruments. The first signs of this market-based approach can be traced back to the 1997 Kyoto Protocol, which introduced the Clean Development Mechanism (CDM), enabling countries to trade emission-reduction credits they had earned through respective projects in developing countries. This early mechanism notwithstanding, most institutions were established after 2007. Important institutions led by public actors are the International Carbon Action Partnership (ICAP) and the Partnership for Market Readiness (PMR). In addition, carbon pricing involves key private institutions, such as the Gold Standard, and public–private cooperation efforts, for instance the Networked Carbon Markets Initiative (NCM).

Having introduced the thematic foci that are at the intersection of both the climate change and energy issue areas, the next sections will discuss the mappings and analyses of the institutional complex governing the climate-energy nexus at the macro level, and the three subfields at the meso level.

3.4 Analyzing the Institutional Complex (Macro Level)

This section presents and discusses our findings on the institutional complex governing the climate-energy nexus at the macro level.[3]

3.4.1 Starting Year

Figure 3.1 shows the increase of institutions addressing climate and energy issues from 1954 to 2016. The dark grey lines represent new institutions per year, while the light grey line shows the cumulative trend.

The oldest institution, and the only one established in the 1950s, is the International Solar Energy Society (ISES), which demonstrates the long history of knowledge on this energy source. Thereafter, it took until the early 1970s for new institutions to emerge and to address the climate-energy nexus. At first glance it seems as if this trend is related to the UN Conference on the Human Environment that took place in 1972 in Stockholm. However, closer inspection shows that these

[3] Please consult Annexes I and II for more detailed information: Annex I provides the complete database of active institutions, including their acronyms and full names along with the data displayed in the figures (starting year, zone, membership, governance functions, and thematic focus); Annex II offers brief descriptions of each institution.

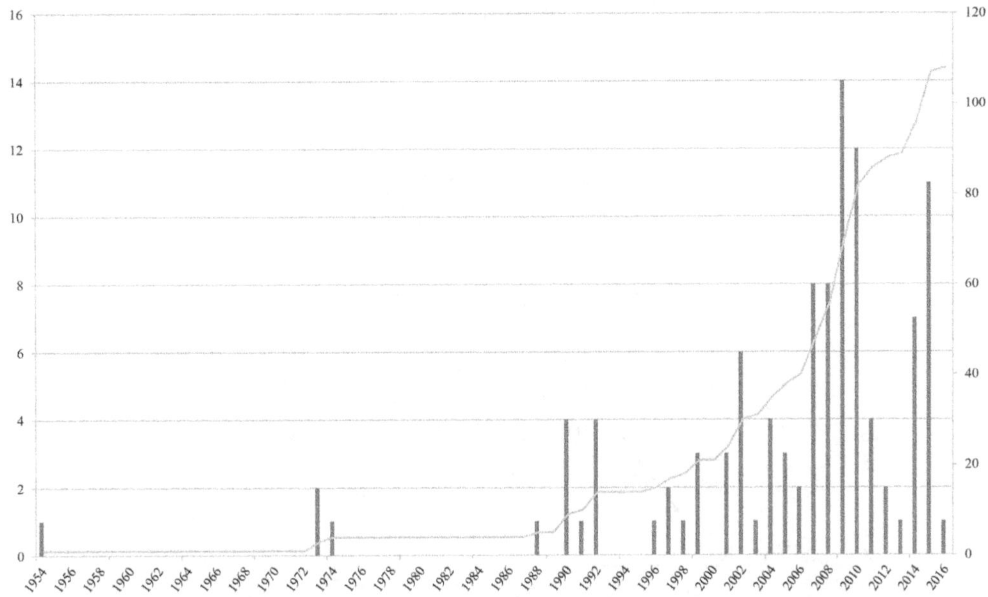

Figure 3.1 Timeline based on starting years of included institutions from 1954 to 2016.

are institutions that were initially established in response to the oil crises in 1970s, which then gradually expanded their activities in light of international climate negotiations in the 1990s. Thereafter, a number of institutions were founded around 1992 when the UN Earth Summit took place in Rio de Janeiro and the UNFCCC was agreed upon. From that point onwards, the graph shows a steep increase in institutions for the following twenty years, and again around 2015, in parallel to the run-up to Agenda 2030 and its SDGs.

3.4.2 Membership

The next subsections are based on membership data. They provide an overview of the overall institutional complex while distinguishing membership types and describing in more detail the distribution and connectedness of institutions and its members.

3.4.2.1 Membership Types per Institution

The institutional complex governing the climate-energy nexus is presented in Figure 3.2. The governance triangle provides insights on the amount of active institutions and the different forms of governance in the institutional complex of the climate-energy nexus. For this constantly changing complex, it provides a

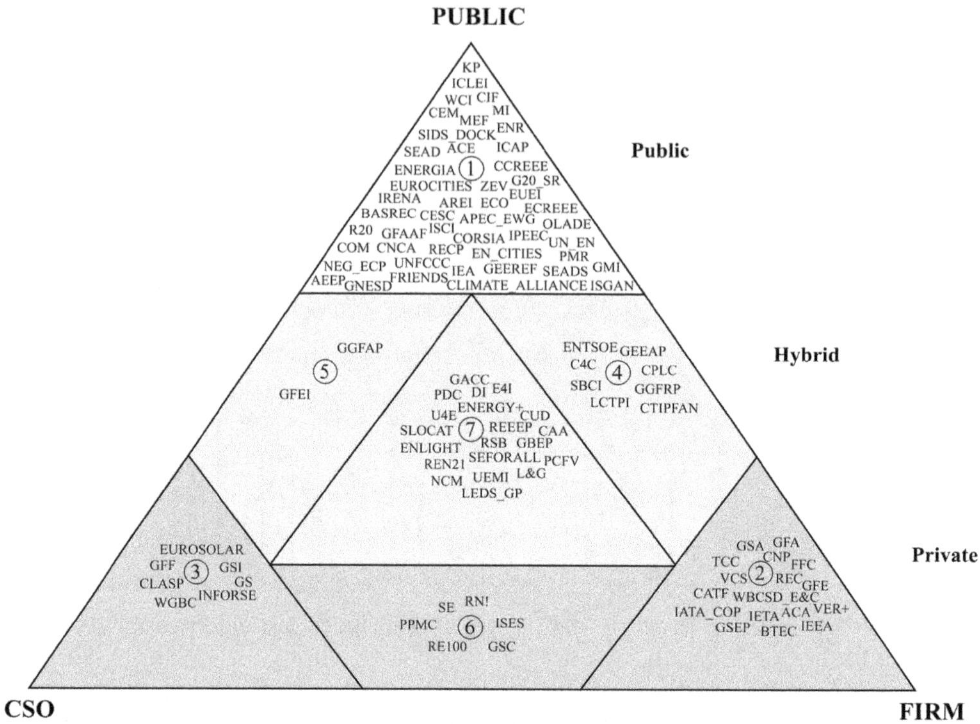

Figure 3.2 Governance triangle of the climate-energy nexus. (Based on Abbott and Snidal 2009a, 2009b, and Abbott 2012; Author's data)

snapshot as of January 2017, when the final cut for mapping was done. The figure structures the different institutions according to their constituting types of actors.

The snapshot of the institutional complex governing the climate-energy nexus comprises 108 institutions. Public agencies are involved in seventy-eight institutions (72 per cent) and exclusively in forty-eight of them (44 per cent) (zone 1). Well-known examples of institutions constituted by solely public actors are the UNFCCC, the Clean Energy Ministerial (CEM), and the IEA. Other, less familiar institutions that fall into zone 1 are, for instance, the Africa Renewable Energy Initiative (AREI) and the International Network on Gender and Sustainable Energy (ENERGIA), which both seek to enhance worldwide access to sustainable energy (AREI 2018; ENERGIA 2018).

Private actors are involved in sixty institutions (56 per cent) and the exclusively private tier (zone 2, 3 and 6) includes a total of thirty institutions (28 per cent). The first core type of private actors, denoted as 'firm' in the figure, i.e. (groups of) firms, investors, and industry associations, are part of fifty-one institutions (47 per cent). Seventeen of these are exclusively constituted by such firm actors, for example the Climate and Energy Cluster of the World Business Council for

Sustainable Development (WBCSD_E&C), which facilitates the sharing of best practices concerning cutting-edge climate and energy topics between its members (WBCSD 2018). The second main type of private actors – NGOs and other organizations representing civil society – are involved in 35 institutions (32 per cent), of which seven are exclusively formed by such CSOs. One example for the latter is the Go Fossil Free (GFF) campaign, committed to a fossil-free society (GFF 2018). Both private actor types, firms and CSOs, cooperate in six institutions, for example in the Global Solar Council (GSC) to promote the uptake of solar energy (GSC 2018).

Public and private actors join forces in the 'hybrid' zones (4, 5, and 7), in which thirty institutions (28 per cent) are situated. These include collaborations between public actors and firms (8), such as the Carbon Pricing Leadership Coalition (CPLC), and cooperative efforts between public actors and CSOs (2), for instance the Global Fuel Economy Initiative (GFEI). This leaves the majority of hybrid institutions (20) to be multi-stakeholder partnerships in which all actor types are included, such as the Global Alliance for Clean Cookstoves (GACC) and Energy for Impact (E4I).

3.4.2.2 Membership Distribution

This subsection examines the membership directories of the individual institutions. Actors are considered members when they have a formal position to influence the rules, norms, operations, and performances of an institution (see Section 3.2.2).[4] The membership data provides insights into the degree of involvement of different actor types across the entire institutional complex as well as per zone in the governance triangle and enables us to explore the level of connectedness between institutions and respective members.

The result of the membership data collection is a total of 13,812 members in the climate-energy nexus (as of January 2017).[5] The number of unique members is 12,241, as one actor can be a member of two or more institutions. There are major differences in the number of members between the institutions; for instance, ICLEI (Local Governments for Sustainability) has 1,156 members, whereas the Western Climate Initiative (WCI) has 'only' 4. Furthermore, there exist differences in the numbers of members per type of actor, as shown in Figure 3.3. Cities are by far the

[4] Discrepancies can exist between the included members and the position of the institution within the governance triangle. For example, the European Network of Transmission System Operators (ENTSOE) is placed in Zone 4. Despite all included members being companies, it is not positioned in Zone 2, because its tasks are stipulated in regulation of the European Commission, a public entity.

[5] Please note that we use an error margin of +-5 per cent for possible data entry mistakes with regard to the included (number of) members. For instance, the member Palau is an island in the Western Pacific Ocean as well as a city in Italy, and can therefore be double coded. We have done our utmost to check for such duplications.

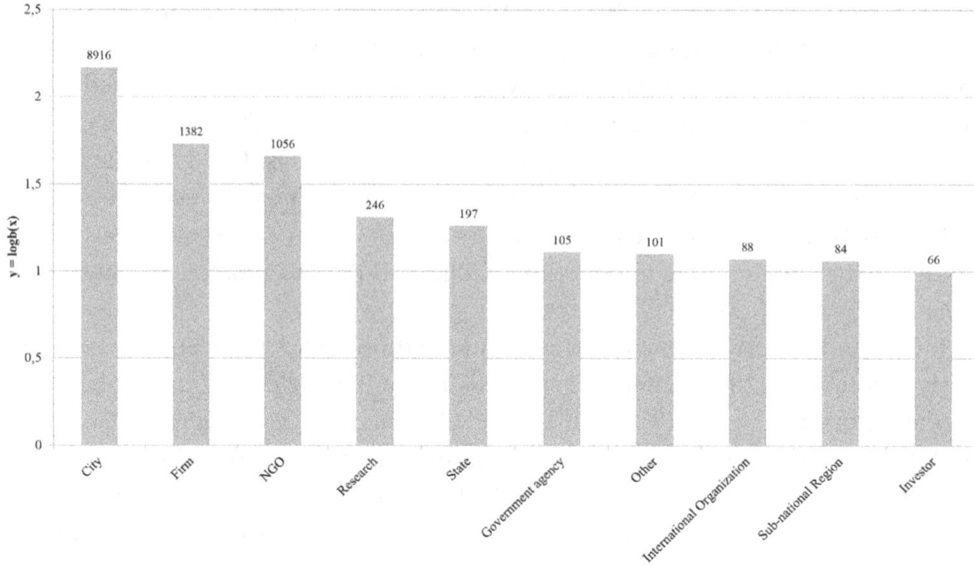

Figure 3.3 Total number of members in different categories (N = 12,241).

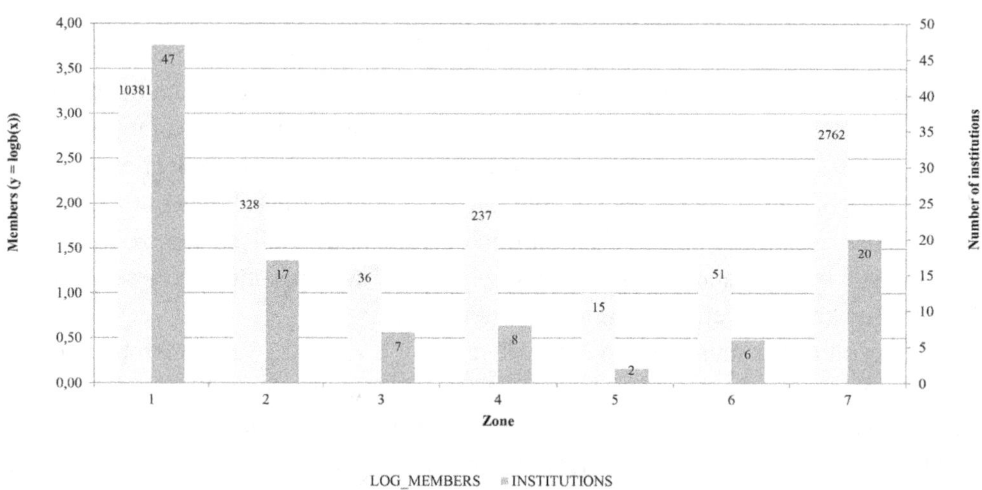

Figure 3.4 Number of members and institutions per zone (1 = public, 2 = firms, 3 = CSOs, 4 = public/firm, 5 = public/CSO, 6 = firm/CSO, 7 = public/firm/CSO).

best-represented type of member, followed by firms and NGOs. The remaining categories are much less prominent.

The distribution of members is reflected in the size of each zone. Figure 3.4 shows the number of members and institutions in each zone. The figure suggests that zone 1 (public) is by far the largest zone, followed by zone 7 (multi-stakeholder), in terms of number of institutions as well as the number of members.

Note that the member categories 'state' and 'government agency' have been merged since it could be argued that government agencies are acting in the name of the state. Moreover, the European Union has been added as it has competency within climate change issues to engage in international treaties on behalf of its member states.

Based on the membership data, it is possible to explore in more detail who the main actors are in addressing the climate-energy nexus. One possibility to depict the centrality of certain actors is a network diagram. We applied this tool for the actor type of states and show the resulting graph in Figure 3.5. The network diagram highlights how the climate-energy nexus consists of a highly

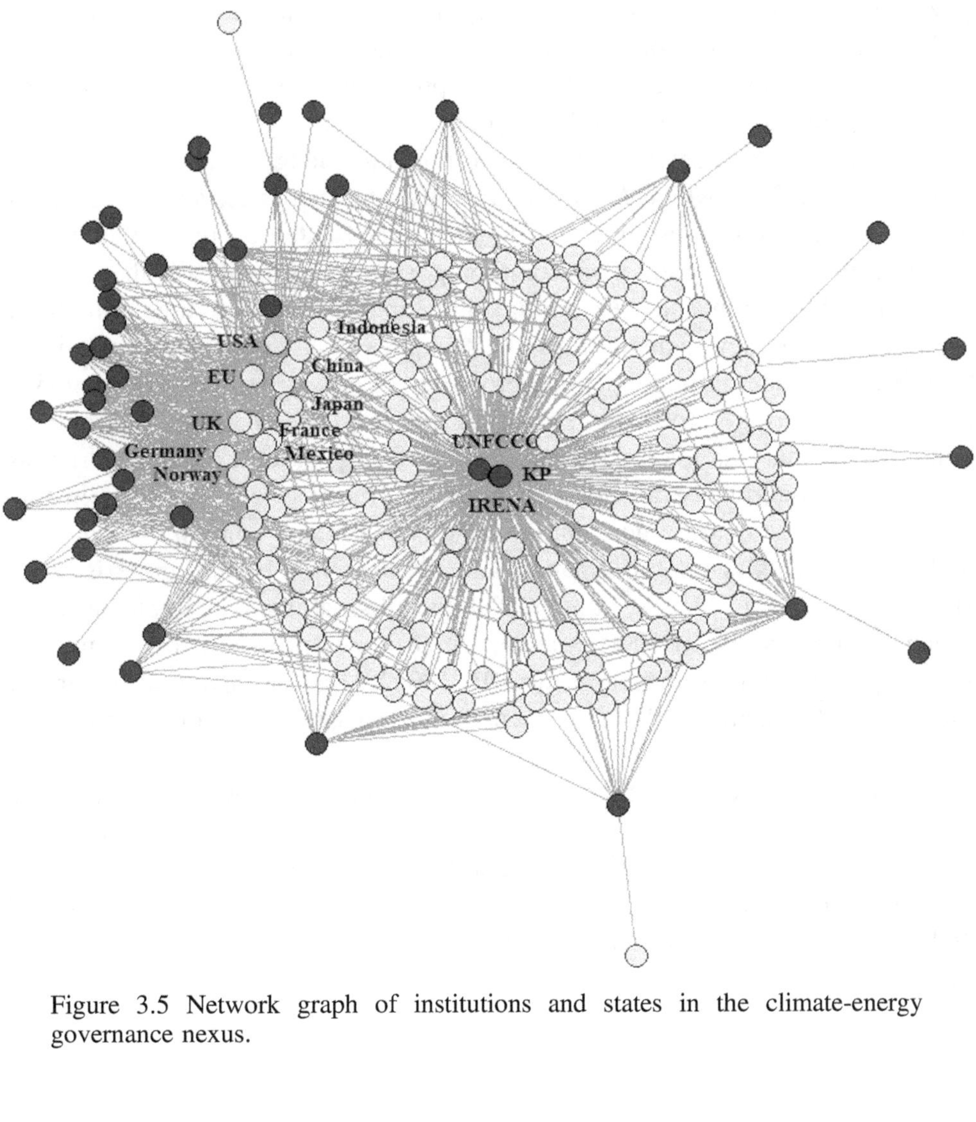

Figure 3.5 Network graph of institutions and states in the climate-energy governance nexus.

interconnected group of actors and institutions. The white nodes represent institutions, whereas the black nodes represent countries. The three central institutions are the UNFCCC, the Kyoto Protocol, and IRENA. The best-connected countries in terms of membership are (in descending order) Germany, the United States, European Union, United Kingdom, Mexico, Norway, China, France, Indonesia and Japan.

3.4.3 Governance Functions

The institutional complex of the climate-energy nexus can also be visualized based on the governance functions of the active institutions. This is displayed in the governance decagon for the climate-energy nexus in Figure 3.6.

'Information and networking' is by far the most common governance function, with thirty-nine occurrences. This implies that most institutions focus on conducting research and publishing reports and/or on organizing meetings and conferences to facilitate the sharing of this knowledge and expertise. One example is IRENA, which serves as a centre of excellence, and a repository of policy, technology, resource, and financial knowledge on renewable energy (IRENA 2018). The second most common governance functions is the combination of 'information and networking' and 'operational' activities, which we found for twenty-seven institutions. This implies that a high number of institutions concentrate on implementing programmes and projects on the ground in addition to sharing information, such as the IEA, which, besides providing authoritative analyses for the full spectrum of energy issues, organizes training and capacity-building workshops (IEA 2018). Furthermore, nineteen institutions focus on 'standards and commitments' and eight are combining 'standards and commitments' with 'information and networking'. Hence, there is a fair share of institutions that seek to introduce rule-making and implementation schemes, such as the RE100 initiative, which unites private actors committed to 100 per cent renewable electricity (RE100 2018). No institution combines the roles of 'standards and commitments' and 'financing', or 'operational' and 'financing'.

Whereas the governance triangle displays a high degree of institutional complexity, particularly in terms of number of institutions, the decagon shows that there is, on top of that, an uneven distribution of the institutions across the governance functions. Table 3.1 summarizes the findings depicted in Figures 3.2 and 3.6, and combines information on membership and governance functions per institution.

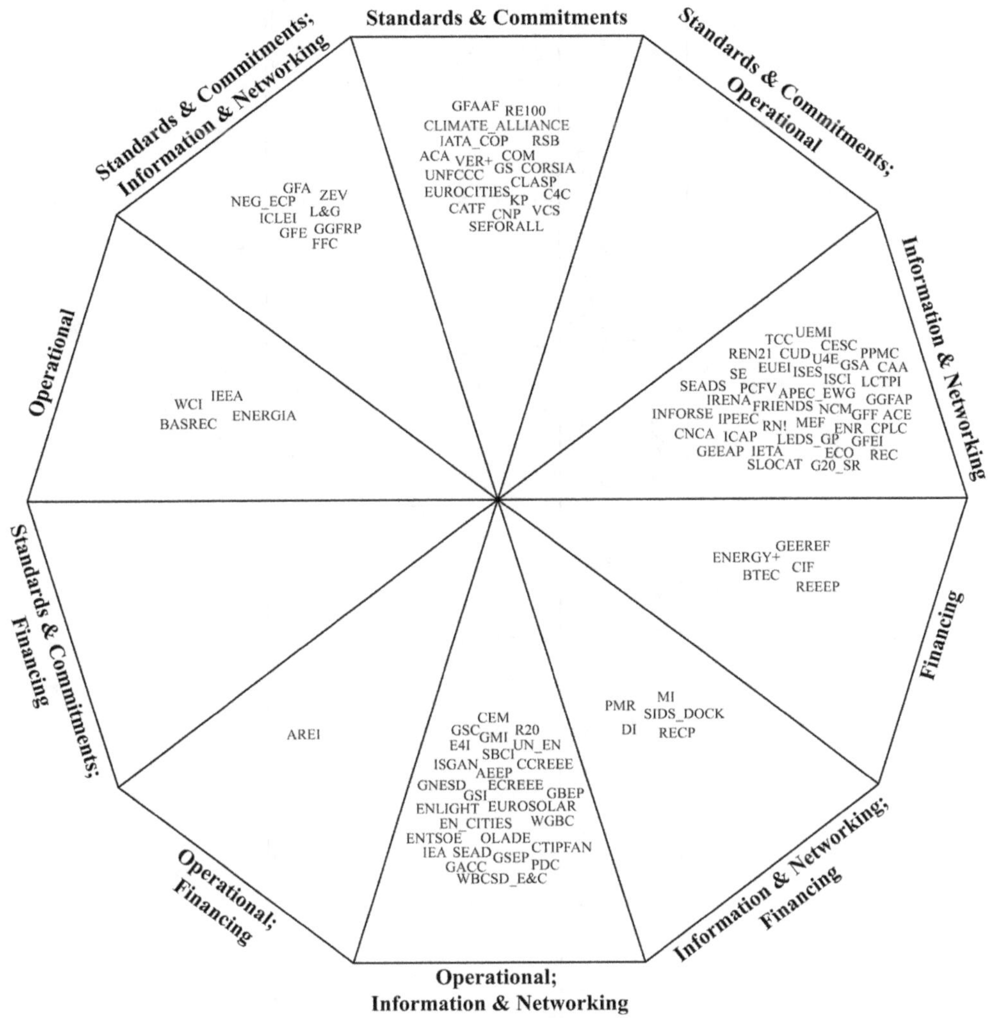

Figure 3.6 Governance decagon of the climate-energy nexus (institutions per governance function).

What can be derived from the table is that the governance function 'standards and commitments' is as much performed by private institutions as by public institutions, while the responsibility for 'information and networking' lies mostly with public institutions. This is in contradiction to the general presumption that public institutions are the ones to set the rules, while private institutions are to perform informal functions, i.e. networking and raising awareness.

Table 3.1 *Total of institutions in the climate-energy nexus per membership and per governance function.*

Zone	Standards & Commitments	Operational	Information & Networking	Financing	Standards & Commitments + Operational	Operational + Information & Networking	Information & Networking + Financing	Standards & Commitments + Information & Networking	Standards & Commitments + Financing	Operational + Financing	Total (Zone)	%
1-public	7	3	15	2	0	13	4	3	0	1	48	44%
2-firm	6	1	4	1	0	2	0	3	0	0	17	16%
3-cso	2	0	2	0	0	3	0	0	0	0	7	6%
4-public/firm	1	0	3	0	0	3	0	1	0	0	8	7%
5-public/cso	0	0	2	0	0	0	0	0	0	0	2	2%
6-firm/cso	1	0	4	0	0	1	0	0	0	0	6	6%
7-public/firm/cso	2	0	9	2	0	5	1	1	0	0	20	19%
Total	19	4	39	5	0	27	5	8	0	1	108	100%

3.4.4 Thematic Focus

Next, we describe the distribution of the institutions in more detail according to their thematic focus by means of the chart in Figure 3.7. The figure highlights important themes at the intersection of climate and energy governance, and the varying degrees of representation, and arguably significance, in terms of institutions.

Most institutions in the climate-energy nexus address several themes and respective goals at the same time to tackle climate and energy challenges. For example, ICLEI, uniting local governments for sustainability, is engaged in energy-efficient city agendas, eco-mobility, and low-carbon development (ICLEI 2018). In addition, many institutions in the institutional complex specifically target increasing the uptake of renewables, such as IRENA and RE100, or transportation modes, such as the Climate Action Takes Flight (CATF) initiative and the Global Green Freight Action Plan (GGFAP). Carbon pricing and trading, and energy efficiency, are the fourth and fifth most preferred themes. Examples of institutions that focus on these two themes are, respectively, CPLC and the International Partnership for Energy Efficiency Cooperation (IPEEC).

As alluded to in Section 3.2.2, the subfields that address these thematic foci should not be seen as silos, but can show overlaps in terms of institutions. First, several institutions focus on 'multiple themes', for instance renewable energy and subsidy reform specifically, and are therefore situated in the two respective subfields. Second, various institutions address a crosscutting theme, for example

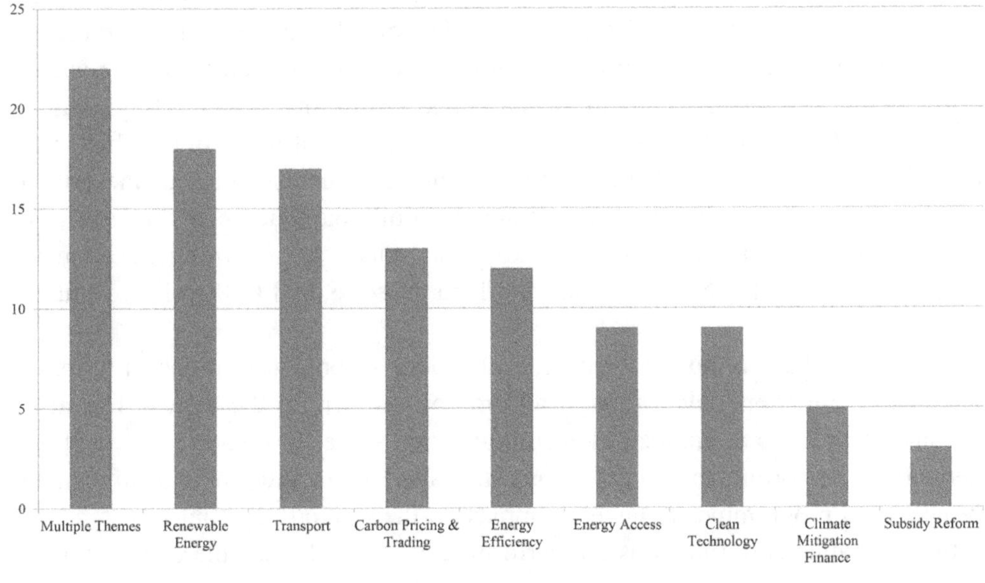

Figure 3.7 Primary Thematic Focus of 108 Institutions.

climate mitigation finance for implementing energy efficiency measures, and are therefore part of both the climate mitigation finance and energy efficiency subfields. Hence, the numbers in Figure 3.7 do not coincide with the mappings of the subfields that are scrutinized in Section 3.5 and the subsequent chapters. However, the figure does provide first insights into the subfields' varying degrees of institutional complexity. Whereas the uptake of renewables is addressed by a high number of institutions, fossil fuel subsidy reform is governed by only a few; meanwhile, the number of institutions on carbon pricing and trading is at the centre of this continuum.

3.4.5 Discussion

Besides the high number of institutions involved in governing the climate-energy nexus, we found a high degree of diversity among these institutions. Institutions constituted by public actors are dominant, although private actors also contribute significantly as the high number of private and multi-stakeholder institutions indicates. Thus, while much scholarly attention has shifted toward transnational global governance, including nonstate and sub-national actors on climate change, energy, and other environmental issues (e.g. Goldthau 2012; Bulkeley et al. 2014; and Hsu et al. 2018), our mapping suggests that state governments and other public agencies retain formal authority in the climate-energy nexus (Jordan et al. 2015).

Analyzing the membership data in more detail further substantiates the dominant role played by (inter)governmental entities. Cities are by far the most represented actor type and, together with other public actors, they occupy more than three-quarters of the entire institutional complex, clearly outmatching the presence of businesses, NGOs, research organizations, investors, and other private actors. Furthermore, the network diagram on states and institutions (figure 3.5) highlights the UNFCCC, Kyoto Protocol, and IRENA as the central institutions which may be explained by the high degree of universality and inclusiveness of these three intergovernmental institutions. The diagram additionally unveils that not only Western (European) countries are of key importance in governing the climate-energy nexus, but that countries such as China, Mexico, and Indonesia are equally involved.

Moreover, there is no clear division of labour among institutions in terms of governance functions, even though all are covered within the nexus. The institutional complex is dominated by institutions that share information and facilitate networking opportunities, and there is a fair share of institutions that implement projects and programmes to have an impact on the ground. By contrast, standard-setting and financing functions are performed by a smaller sample of institutions. These findings illustrate that, within the climate-energy nexus, private institutions do not shy away from setting standards, while public institutions predominantly perform informal functions. Further studies have to show whether this uneven distribution of governance functions has implications for the performance of the institutional complex of the nexus as a whole, i.e. at the macro level.

Furthermore, the examination of key themes at the intersection of climate and energy governance provides interesting first insights into the subfields of the climate-energy nexus. First and foremost, there is a high number of institutions focusing on multiple themes, which suggests that there are many overlapping institutions among the subfields, an issue that will be explored in subsequent chapters. Second, the degree of institutional complexity appears to vary across the subfields. The numbers of institutions addressing the respective thematic foci suggest that the subfield for renewable energy is most densely populated, whereas the subfield for fossil fuel subsidy reform consists of only a few institutions. Given that carbon pricing is in the middle of this continuum, the three subfields covered in the book provide a useful variation for the analyses in Chapters 4–6 and the comparative study in Chapter 8.

In sum, over the past twenty-five years the institutions governing the climate-energy nexus evolved into a densely populated complex dominated mostly by public actors, in which governance functions are unevenly spread, and the degree of complexity varies considerably across the subfields.

3.5 Analyzing the Subfields: Renewable Energy, Fossil Fuel Subsidy Reform, and Carbon Pricing (Meso Level)

This section describes and analyzes those institutions that, exclusively or as one of their multiple foci, address one of the three subfields of the climate-energy nexus covered in the book. To illustrate the three meso-level institutional complexes, similar figures are used as in the previous section: the governance triangle providing insights on the types of actors involved, and the governance decagon displaying the governance functions of the individual institutions. Like in the previous section, the illustrations are snapshots of the subfields as of January 2017.

3.5.1 Renewable Energy

Figure 3.8 presents the governance triangle for the institutions that aim to promote the uptake of renewables globally. The institutional complex comprises forty-six institutions, making the renewable energy subfield the largest within the climate-energy nexus. Nineteen of these institutions focus exclusively on renewable energy, while for the remaining number renewables are but one part of their portfolio.

Most institutions within the governance triangle are constituted by public actors, including (groups of) states, international organizations, cities, and regions. These actors are involved in thirty-eight institutions, of which twenty-eight are purely public (zone 1). Widely known examples of the latter are IRENA and the IEA,

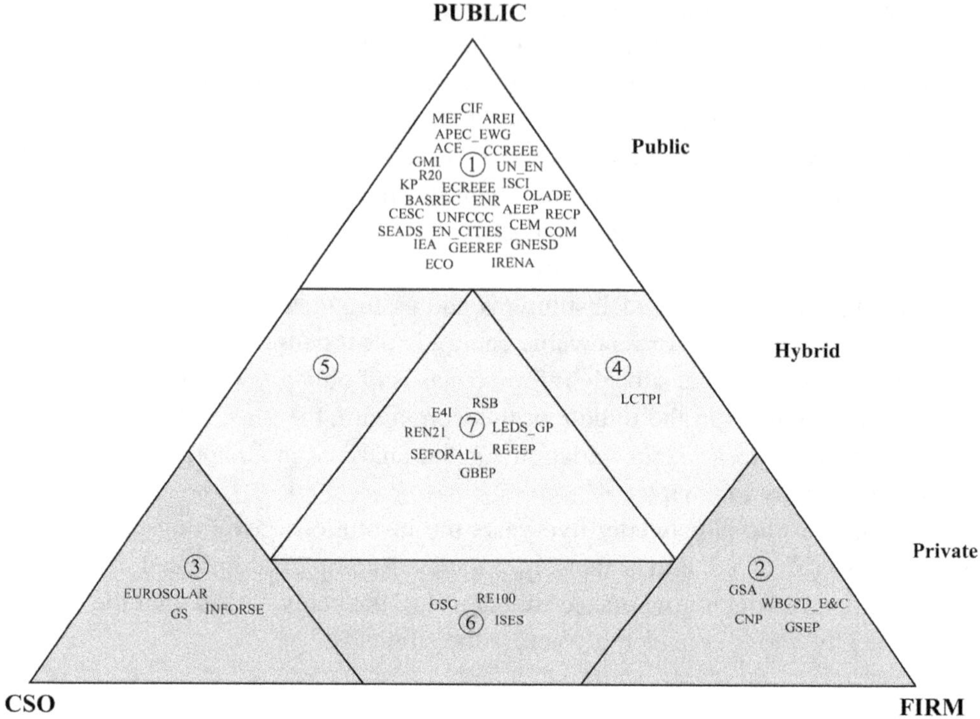

Figure 3.8 Governance Triangle for the subfield of renewable energy. (Based on Abbott and Snidal 2009a; 2009b; Abbott 2012; and Author's data)

while less familiar ones are, for instance, Energy Cities (EN_CITIES) and the International Solar Cities Initiative (ISCI), both involving local authorities committed to facilitating energy transitions.

Private actors engage in eighteen institutions while the exclusively private tier (zone 2, 3, and 6) counts ten institutions. (Groups of) firms, investors, and industry associations take part in seventeen institutions (37 per cent), of which four are purely firm-based. These include, among others, the Carbon Neutral Protocol (CNP), which provides instruments to achieve carbon neutrality such as renewable energy certificates (CNP 2018). The third main category of actors, NGOs and other CSOs, is involved in thirteen institutions (28 per cent), of which two exclusively include CSOs. One example is the International Network for Sustainable Energy (INFORSE), which constitutes a network of NGOs working on sustainable energy solutions (INFORSE 2018). On top of that, different types of firm and CSO actors join forces in three institutions: GSC, RE100, and ISES.

Public and private actors unite in the 'hybrid' zones (zones 4, 5, and 7) through eight institutions (17 per cent), of which seven are multi-stakeholder partnerships, including all three types of actors. Well-known multi-stakeholder partnerships are Sustainable Energy for All (SEFORALL) and REN21, which are committed to, respectively, substantially increasing the uptake of renewables by 2030 in

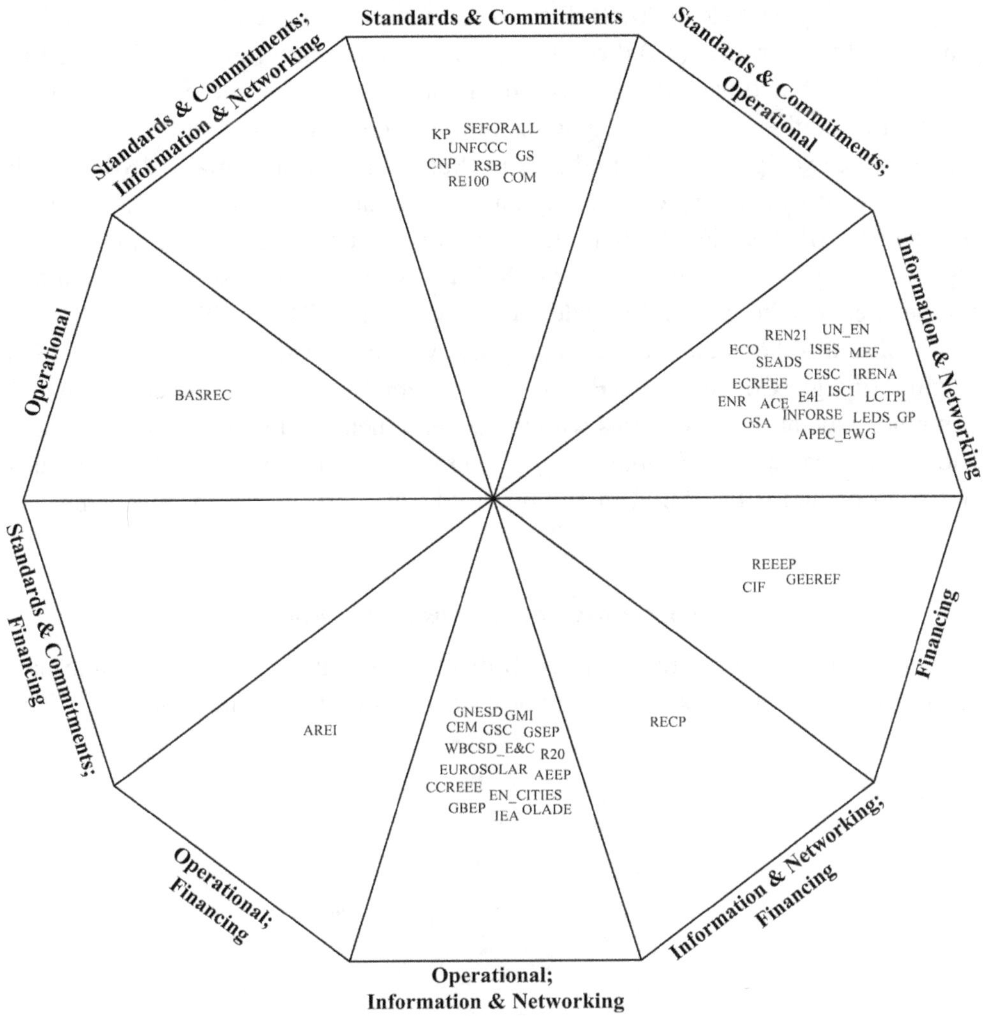

Figure 3.9 Governance decagon for the sub-field of renewable energy (institutions per governance function) (Author's data).

accordance to the SDG 7 target, and to connecting stakeholders to facilitate joint action toward a global transition to renewables (SEforALL 2018; REN21 2018). Finally, the one institution in which public actors partner up with firm-based actors is the Low Carbon Technology Partnerships initiative (LCTPI). This is a partnership between the WBCSD, the IEA, and the UN's Sustainable Development Solutions Network. Its REscale programme brings together energy and technology companies who aim to accelerate the deployment of renewables (LCTPi 2018).

The governance decagon in Figure 3.9 highlights the governance functions of the individual institutions. Most of them (17) govern through 'information and networking' and 'operational' activities simultaneously, for example Regions for Climate Action (R20), which combines informing and supporting climate-resilient project development (R20 2018). Fifteen institutions solely concentrate on 'information and networking' (15). For instance, it is the primary function of REN21 to inform the international community on the status of renewable energy (REN21 2018). In addition, there are eight institutions that set 'standards and commitments'. One example is the Roundtable for Sustainable Biomaterials (RSB), which provides a certification scheme supporting the development of a sustainable bioeconomy (RSB 2018). Only three institutions focus on developing and providing 'financing' mechanisms, including REEEP, which aims to strengthen markets for clean energy in low- and middle-income countries (REEEP 2018).

Finally, all other fields are at best filled with one institution. The Baltic Sea Region Energy Cooperation (BASREC) prefers 'operational' practices, AREI combines 'operational' activities with the development of 'financing' mechanisms, and the Africa-EU Renewable Energy Cooperation Program (RECP) facilitates 'information and networking' in addition to developing 'financing' mechanisms.

3.5.2 Fossil Fuel Subsidy Reform

The governance triangle in Figure 3.10 displays which types of actors take part in institutions that work toward removing harmful fossil fuel subsidies. The

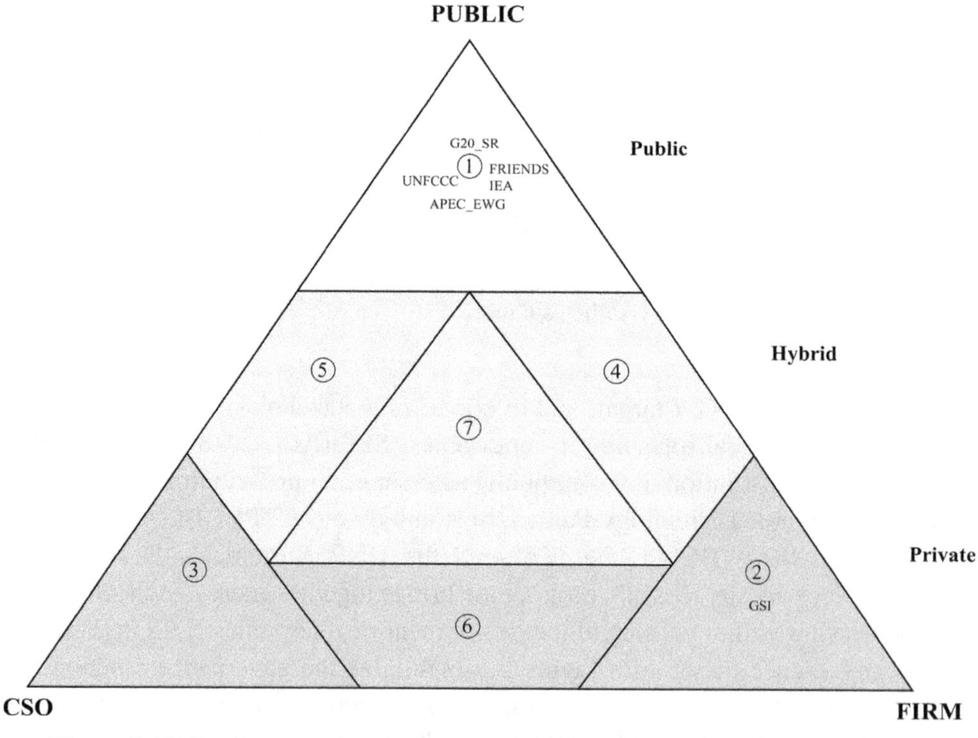

Figure 3.10 Governance triangle for the sub-field of fossil fuel subsidy reform. (Based on Abbott and Snidal 2009a; 2009b; Abbott 2012; and Author's data)

institutional complex comprises only a small number of institutions (six), with half of these addressing subsidy reform as their main priority.

Clearly, the largest share of institutions (five) has exclusively public membership, including some of the leading institutions in the climate-energy nexus: the UNFCCC, Friends, IEA, the G20 Subsidy Reform, and the APEC Energy Working Group. Only one institution is constituted by firms, namely GSI. This initiative is led by the International Institute for Sustainable Development, and supports international processes, national governments, and CSOs to align subsidies with sustainable development (IISD 2018).

The governance functions of the six institutions are visualized in Figure 3.11. It shows that 'information and networking' is the predominant way in which institutions

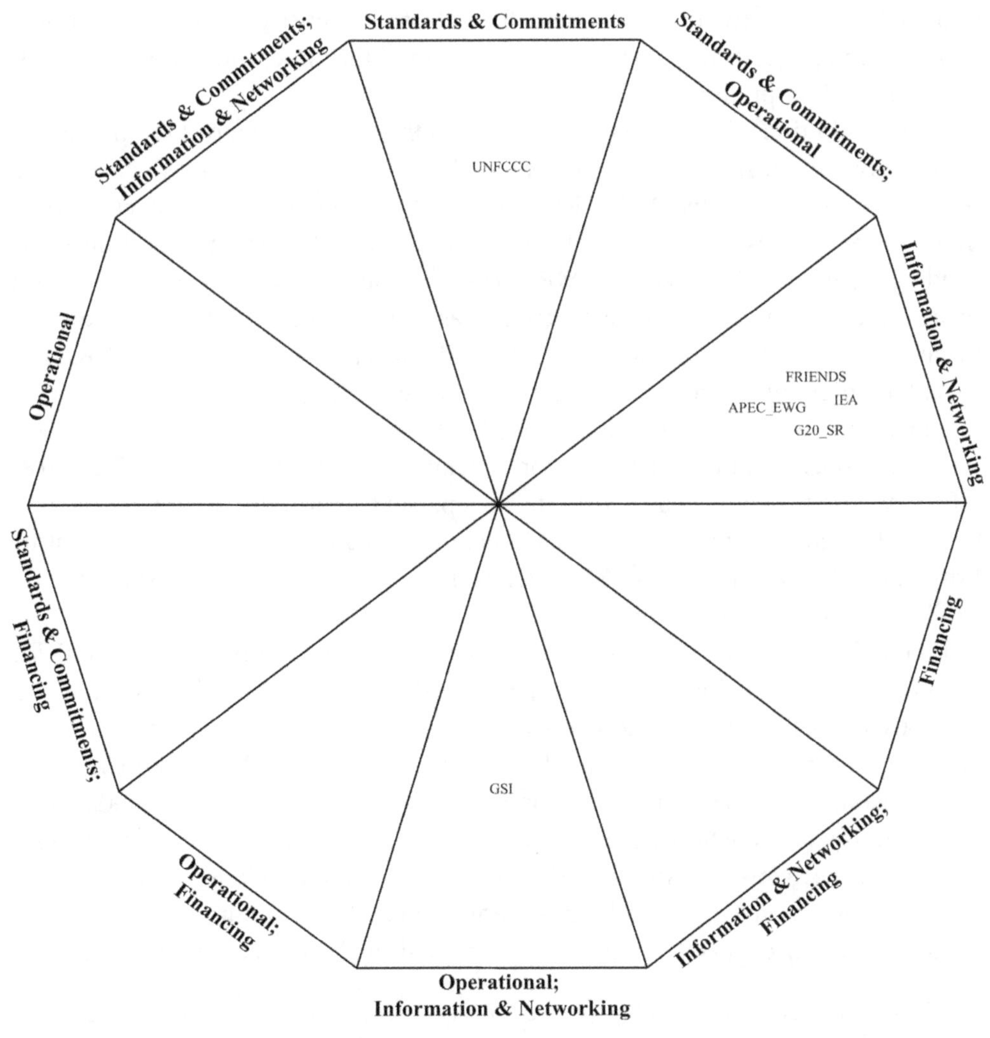

Figure 3.11 Governance decagon for the fossil fuel Subsidy reform sub-field (institutions per governance function) (Author's data).

on fossil fuel subsidy reform operate. No less than four of the six institutions adhere to this governance function. These include the IEA, Friends, APEC Energy Working Group, and G20 Subsidy Reform, which broadly focus on sharing information and organizing events to convince their members and the wider community of the urgency to reform fossil fuel subsidies. The remaining two institutions concentrate on standard-setting for this subfield (UNFCCC) and, respectively, 'information and networking' in combination with 'operational' activities (GSI).

3.5.3 Carbon Pricing

The governance triangle in Figure 3.12 includes all institutions in the climate-energy nexus dataset that aim at putting a price on carbon, facilitating a carbon trade scheme, or providing a system for offsetting emissions. The institutional complex comprises fifteen individual institutions.[6] Of these, thirteen focus exclusively on carbon pricing.

Most institutions fall exclusively either into the public arena or the firm sector. Concretely, six institutions are constituted by (groups of) states, cities, and regions, for example, the WCI that supports the implementation of emissions-trading programmes, and PMR that offers country-specific guidance on Emissions Trading Registries (WCI 2018; PMR 2018). Six others count solely (groups of) firms, investors, and industry associations as their members. These include, for instance, the Verified Carbon Standard (VCS, recently renamed to VERRA), a voluntary programme for certifying emission-reduction projects, and the International Emissions Trading Association (IETA), which aims at establishing effective market-based trading systems for greenhouse gas emissions (IETA 2018; VERRA 2018). Finally, three institutions represent hybrid efforts. These include the UN Global Compact Caring for Climate (C4C) and CPLC, which consist of firms and public actors, and NCM, which is the only institution in this subfield that exhibits all three types of actors.

Figure 3.13 presents the governance decagon for the carbon pricing subfield, depicting the governance functions performed by each institution. Unlike the previous figures for renewable energy and fossil fuel subsidy reform, the main governance function for this subfield is setting 'standards and commitments' (by seven institutions). While the VCS and the CNP have established standards to certify emission reductions, the International Air Transport Association (IATA) provides a mechanism to offset carbon emissions for the aviation sector (IATA 2018). 'Information and networking' is performed by four institutions, for example by ICAP, which connects countries, subnational, and supranational entities that

[6] VER+ is, at the time of writing, no longer operative, but still included in the dataset, for which the final cut was made in January 2017.

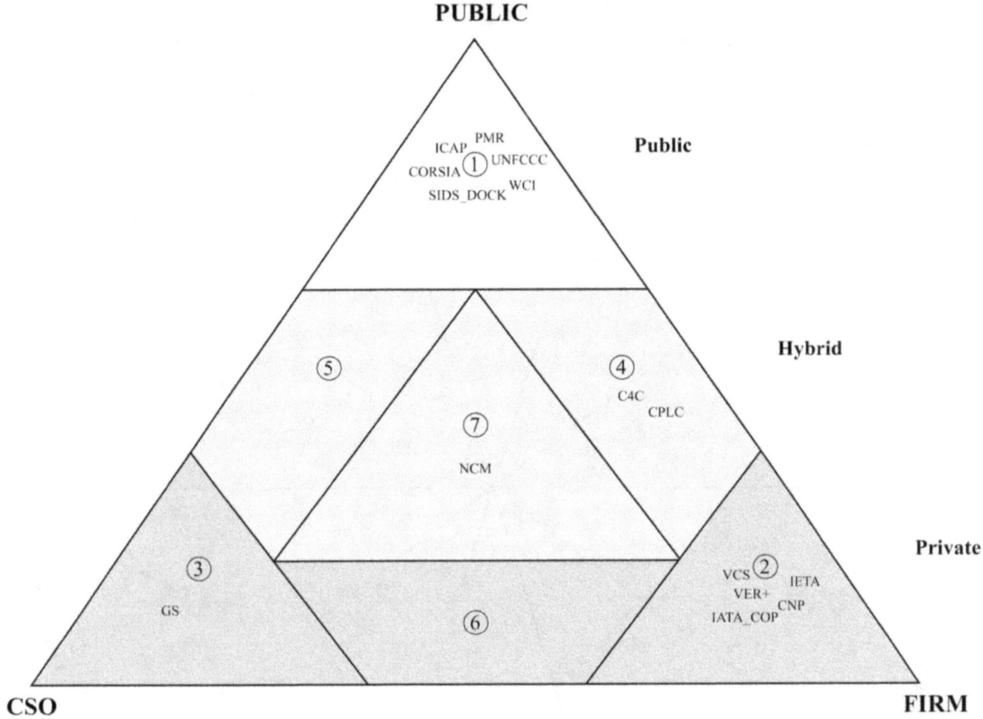

Figure 3.12 Governance triangle for the sub-field of carbon pricing. (Based on Abbott and Snidal 2009a; 2009b; Abbott 2012; and Author's data)

either have established carbon markets or plan to do so (ICAP 2018). In addition, PMR and the Small Island Developing States Sustainable Energy Initiative (SIDS_DOCK) combine 'information and networking' with the development of 'financing' mechanisms. Finally, WCI falls into the category 'operational' as it provides administrative and technical services supporting the implementation of emissions-trading programmes.

3.5.4 Discussion

Having scrutinized the three subfields for renewable energy, fossil fuel subsidy reform, and carbon pricing, we arrive at interesting first insights on each of the institutional complexes. One key observation concerns regards the varying degrees of institutional complexity across the subfields.

The highest degree of complexity was found for the renewable energy subfield, in terms of the number of institutions, but also, as this section has shown, regarding memberships and governance functions. The subfield includes a wide range of public and private actors, while public institutions dominate. All governance

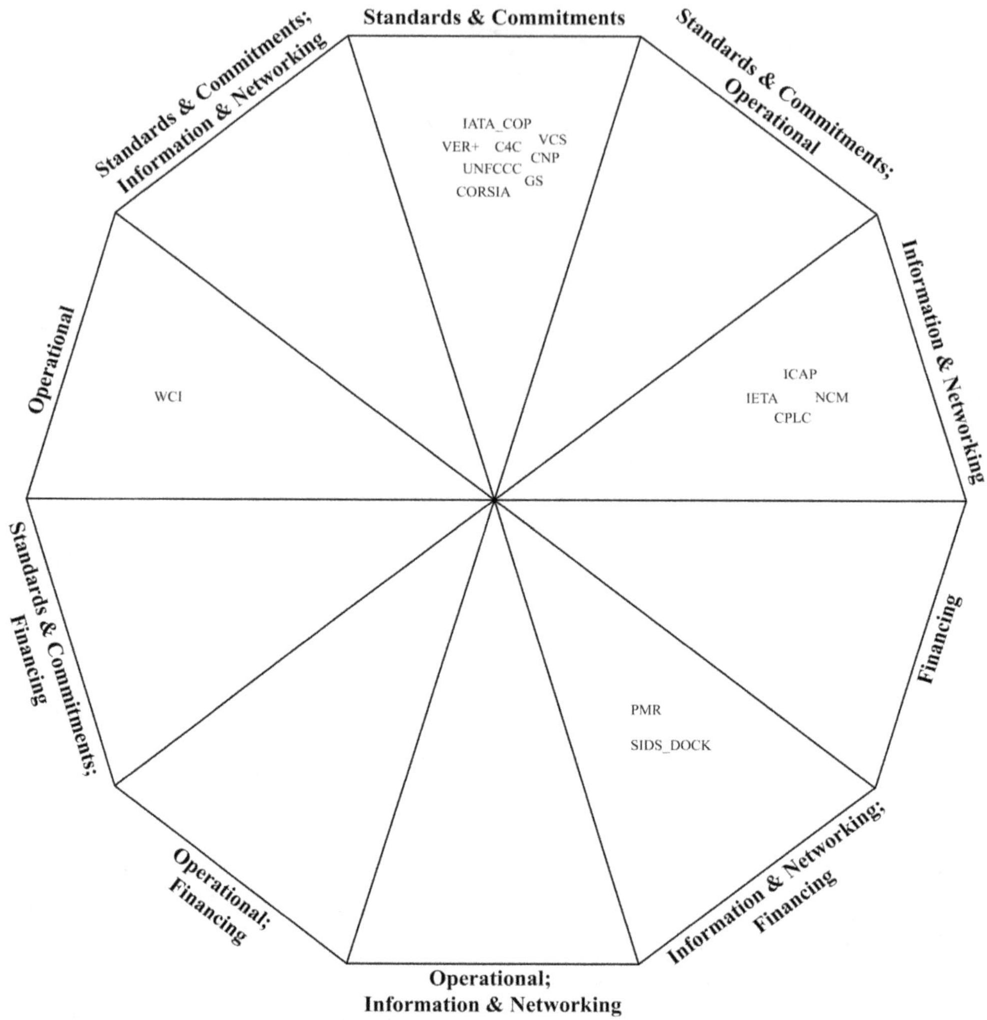

Figure 3.13 Governance decagon for the subfield of carbon pricing (institutions per governance function) (Author's data).

functions are addressed, while the main emphasis is laid on information-sharing, networking, and operational activities. Moreover, the subfield shows characteristics similar to the overall institutional complex addressing the climate-energy nexus – which is clearly related to the fact that the subfield of renewable energy is constituted by almost half of the institutions in the entire dataset.

The lowest degree of complexity, in turn, was found in the fossil fuel subsidy reform subfield: there are only seven institutions, of which six are exclusively public, and information and networking is the dominant instrument to govern subsidy reform. This said, it is important to consider other institutions as well.

These have not been included in this chapter's dataset, since they do not fit our selection criteria (Section 3.2.1). This is mostly due to their chief thematic foci that divert from, or go beyond, the climate and energy domains. Examples are the Organisation for Economic Co-operation and Development (OECD), the World Trade Organization (WTO), and the IMF. To complete the picture, these institutions are included and elaborated on in Chapter 5, which provides an in-depth analysis of the subfield.

In the middle of the continuum is the carbon pricing subfield, not solely with regard to the number and types of institutions involved, but also in terms of the governance functions they perform. The subfield features a balance of exclusively public institutions and purely private institutions. Furthermore, carbon pricing is predominantly governed through setting standards and commitment. This mostly includes certification schemes for emission reductions and carbon offsetting. Finally, we found an important distinction between institutions that aim to establish prerequisites for a carbon market and those that focus on existing ones. For instance, SIDS_DOCK, an alliance between small island states, assists its members in connecting national energy sectors to EU and US carbon markets, but it does not facilitate carbon pricing, trading, or offsetting itself. As a result, this creates a discrepancy between the dataset included in this chapter and the cases studied in Chapter 6.

Furthermore, it is of interest to examine overlapping institutions across the three subfields and to what extent these represent thematic overlaps. Fossil fuel subsidy reform inherently supports the uptake of renewable energy; when harmful subsidies are removed, fossil fuels make way for renewables. Nevertheless, the fossil fuel subsidy reform and renewable energy subfields share only two institutions: the APEC Energy Working Group and the IEA, which are both intergovernmental cooperation efforts addressing energy issues in the broadest sense. Similarly, while putting a price on carbon provides an incentive to shift to renewable energy, only the CNP is an overlapping institution bridging both subfields. Finally, the one institution shared by all three subfields, and therewith providing the only connection between the fossil fuel subsidy reform and carbon pricing subfields, is the UNFCCC. This suggests that the central role of the UNFCCC regime is not restricted to global climate change governance, but transcends well into global energy governance as a major hub targeting the low-carbon transformation of energy systems.

3.6 Conclusions

The Paris Agreement and Agenda 2030, and particularly SDG 7 to ensure sustainable energy for all, highlight the importance of an effective and integrated approach to the interconnected climate and energy challenges. While existing studies have provided insights into the institutional complexity of either global climate or energy governance, governance scholars had yet to uncover the institutional complex addressing the nexus between both domains. Against this backdrop, this chapter first identified and mapped the macro level of this nexus, i.e. all institutions that address climate and energy challenges simultaneously. Furthermore, the chapter zoomed in on the institutions that constitute three subfields at the meso level: renewable energy, fossil fuel subsidy reform, and carbon pricing.

The climate-energy nexus can be characterised as institutionally complex, reflecting the past twenty-five years of intense discussions in the international community on both challenges and sustainable development more broadly. This chapter conveyed and analyzed this complexity in various ways. The climate-energy nexus involves a high number of very different institutions – with a strong presence of exclusively public institutions, while private actors contribute significantly through a fair share of private and multi-stakeholder institutions. Membership data further substantiates the dominance of public actors, and cities particularly. Second, even though all governance functions are performed across the institutional complex, there appears to be a strong focus on information and networking, and operational activities to some extent, while only a minority of institutions in the nexus focus on standards and commitments and financing mechanisms. Finally, while most institutions target various themes and activities, most attention is directed toward the uptake of renewables, transforming the transport sector, carbon pricing, and increasing energy efficiency.

Delving into the selected subfields at the meso level provided first insights on the differences and commonalities across the three respective complexes. The renewable energy subfield has the highest concentration of public, private, and hybrid institutions, which focus mostly on information-sharing and networking. On the other end, the subfield of fossil fuel subsidy reform is sparsely populated and marked by a prevalence of public institutions, which predominantly govern through information and networking. This leaves the carbon pricing subfield at the centre of the continuum, exhibiting a balance between public and private institutions, with both camps focusing mostly on standards and commitments.

While this chapter has provided an innovative and novel mapping, it could not go into detail on the processes taking place within and across the institutional complexes. Such much-needed analyses on questions of coherence, management, legitimacy, and effectiveness in the climate-energy nexus – will be provided by the subsequent chapters of this book. The mapping serves as the empirical basis for these crucial endeavours. First, the comprehensive dataset has helped to select appropriate case studies for Chapters 4–6, which explore varying degrees of coherence and management attempts within the three subfields. Second, our chapter provides key data for evaluating the effectiveness and legitimacy of five individual institutions in Chapter 7. Finally, our findings support the comparative assessment of institutional effectiveness for the three subfields in Chapter 8, from which lessons can be drawn for the performance of the overall institutional complex governing the climate-energy nexus.

By the same token, the dataset presented in this chapter is the first step toward creating a knowledge base that can serve as a tool for policy makers, businesses, and other organizations alike. It improves our understanding of the institutional complexity that characterizes the climate-energy nexus, and guides actors to navigate the institutionally complex global climate and energy governance landscape.

3.7 References

Abbott, K. W. 2012. The Transnational Regime Complex for Climate Change. *Environment and Planning C: Government and Policy* 30(4), 571–590. https://doi.org/10.1068/c11127.

Abbott, K. W. and Snidal, D. 2009a. Strengthening International Regulation through Transnational New Governance: Overcoming the Orchestration Deficit. *Vanderbilt Journal of Transnational Law* 42(2), 501–578.

Abbott, K. W. and Snidal, D. 2009b. The Governance Triangle: Regulatory Standards Institutions and the Shadow of the State. *The Politics of Global Regulation*, 44–88.

APEC (Asia-Pacific Economic Cooperation). 2009. Leaders' Declaration. Available at www.apec.org/Meeting-Papers/Leaders-Declarations/2009/2009_aelm.

AREI (Africa Renewable Energy Initiative). 2018. About AREI. Available at www.arei.org/#about.

Bäckstrand, K. 2008. Accountability of Networked Climate Governance: The Rise of Transnational Climate Partnerships. *Global Environmental Politics* 8(3), 74–102. https://doi.org/10.1162/glep.2008.8.3.74.

Barnsley, I. and Ahn, S. 2014. *Mapping Multilateral Collaboration on Low-Carbon Energy Technologies*. Paris. Available at www.iea.org/publications/insights/insightpublications/MappingMultilateralCollaboration_FINAL.pdf.

Biermann, F., Pattberg, P., van Asselt, H., and Zelli. F. 2009. The Fragmentation of Global Governance Architectures: A Framework for Analysis. *Global Environmental Politics* 9(4), 14–40. https://doi.org/10.1162/glep.2009.9.4.14.

Bulkeley, H., Andonova, L., Betsill, M. M., Compagnon, D., Hale, T., Hoffmann, M. J., Newell, P., Paterson, M., Roger, C., and VanDeveer, S. D. 2014. *Transnational Climate Change Governance*. Cambridge: Cambridge University Press. https://doi.org/10.1017/CBO9781107706033.

Burniaux, J. and Chateau, J. 2014. Greenhouse Gases Mitigation Potential and Economic Efficiency of Phasing-Out Fossil Fuel Subsidies. *International Economics* 140(573), 71–88. https://doi.org/10.1016/j.inteco.2014.05.002.

CNP (Carbon Neutral Protocol). 2018. Carbon Neutral. Available at www.carbonneutral.com.

Colgan, J. D., Keohane, R. O., and Van de Graaf, T. 2012. Punctuated Equilibrium in the Energy Regime Complex. *Review of International Organizations* 7, 117–143. https://doi.org/10.1007/s11558-011-9130-9.

Colgan, J. D. and Van de Graaf, T. 2014. Mechanisms of Informal Governance: Evidence from the IEA. *Journal of International Relations and Development* 18(4), 455–481.

Dubash, N. K. and Florini, A. 2011. Mapping Global Energy Governance. *Global Policy* 2, 6–18. https://doi.org/10.1111/j.1758-5899.2011.00119.x.

ENERGIA (International Network on Gender and Sustainable Energy). 2018. International Network on Gender and Sustainable Energy. Available at www.energia.org/.

Escribano, G. 2015. Fragmented Energy Governance and the Provision of Global Public Goods. *Global Policy* 6(2), 97–106. https://doi.org/10.1111/1758-5899.12195.

EUROSOLAR. 2017. Replacing Nuclear and Fossil Fuels Entirely with Renewable Energy Sources. Annual Report 2016.

Falkner, R. 2014. Global Environmental Politics and Energy: Mapping the Research Agenda. *Energy Research and Social Science* 1, 188–197. https://doi.org/10.1016/j.erss.2014.03.008.

Florini, A. and Sovacool, B. K. 2009. Who Governs Energy? The Challenges Facing Global Energy Governance. *Energy Policy* 37(12), 5239–5248. https://doi.org/10.1016/j.enpol.2009.07.039.

G20 (Group of 20). 2009. Leaders' Statement. Available at www.g20.utoronto.ca/2009/2009communnique0925.html.

GFF (Go Fossil Free). 2018. Go Fossil Free. Available at https://gofossilfree.org/.

Goldthau, A. 2012. From the State to the Market and Back: Policy Implications of Changing Energy Paradigms. *Global Policy* 3(2), 198–210. https://doi.org/10.1111/j.1758-5899.2011.00145.x.

Greenpeace. 2008. *False Hope: Why Carbon Capture and Storage Won't Save the Climate*. Amsterdam. Available at www.greenpeace.org/usa/wp-content/uploads/legacy/Global/usa/report/2008/5/false-hope-why-carbon-capture.pdf.

GSC (Global Solar Council). 2018. Global Solar Council. Available at www.globalsolarcouncil.org/.

Hale, T. and Roger, C. 2014. Orchestration and Transnational Climate Governance. *The Review of International Organizations* 9(1), 59–82. https://doi.org/10.1007/s11558-013-9174-0.

Hoffmann, Matthew J. 2011. *Climate Governance at the Crossroads. A Global Response after Kyoto*. Oxford: Oxford University Press. https://doi.org/10.1093/acprof:oso/9780195390087.001.0001.

Hsu, A., Widerberg, O., Weinfurter, A., Chan, S., Roelfsema, M., Lütkehermöller, K., and Bakhtiari, F. 2018. Bridging the Emissions Gap – The Role of Non-State and Subnational Actors. In *The Emissions Gap Report 2018. A UN Environment Synthesis Report*. Nairobi: United Nations Environment Programme.

IATA (International Air Transport Association). 2018. Carbon Offset Program. Available at www.iata.org/whatwedo/environment/Pages/carbon-offset.aspx.

ICAP (International Carbon Action Partnership). 2018. About ICAP. Available at https://icapcarbonaction.com/en/partnership/about.

ICLEI (International Council for Local Environmental Initiatives). 2018. ICLEI Local Governments for Sustainability. Available at www.iclei.org/en/Home.html.

IEA (International Energy Authority). 2018. IEA. Our Mission. Available at www.iea.org/about/ourmission/.

IETA (International Emissions Trading Association). 2018. Our Mission. Available at www.ieta.org/Our-Mission.

IISD (International Institute for Sustainable Development). 2018. About GSI. Available at www.iisd.org/gsi/about.

INFORSE (International Network for Sustainable Energy). 2018. International Network for Sustainable Energy. Available at www.inforse.org/.

IPCC (Interngovernmental Panel on Climate Change). 2018. Summary for Policymakers. In *Global Warming of 1.5°C. An IPCC Special Report on the Impacts of Global Warming of 1.5°C above Pre-industrial Levels and Related Global Greenhouse Gas Emission Pathways, in the Context of Strengthening the Global Response to the Threat of Climate Change*, edited by V. Masson-Delmotte, P. Zhai, H. O. Pörtner, D. Roberts, J. Skea, P. R. Shukla, A. Pirani, et al. Geneva: World Meteorological Organization.

IRENA (International Renewable Energy Agency). 2018. About IRENA. Available at irena.org/aboutirena.

Jordan, A. J., Huitema, D., Hildén, M., van Asselt, H., Rayner, T. J., Schoenefeld, J. J., Tosun, J., Forster, J., and Boasson, E. L. 2015. *Nature Climate Change* 5, 977–952.

Keohane, Robert O. and Victor, David G. 2011. The Regime Complex for Climate Change. *Perspectives on Politics* 9(1), 7–23. https://doi.org/10.1017/S1537592710004068.

Kérébel, C. and Keppler, Jan Horst. 2009. *La Gouvernance Mondiale de L'Énergie.* Gouvernance Européenne et Géopolitique de L'Énergie. Paris & Bruxelles.

LCTPi (Low Carbon Technology Partnerships initiative). 2018. REscale – Scaling up Renewables. Available at https://lctpi.wbcsd.org/portfolio-item/renewables/.

Leal-Arcas, Rafael and Filis, Andrew. 2013. The Fragmented Governance of the Global Energy Economy: A Legal-Institutional Analysis. *Journal of World Energy Law and Business* 6(4), 348–405. https://doi.org/10.1093/jwelb/jwt011.

Lesage, D., Van de Graaf, T., and Westphal, K. 2010. *Global Energy Governance in a Multipolar World.* Farnham & Burlington: Ashgate Publishing.

Pattberg, P. and Stripple, J. 2008. Beyond the Public and Private Divide: Remapping Transnational Climate Governance in the 21st Century. *International Environmental Agreements: Politics, Law and Economics* 8(4), 367–388. https://doi.org/10.1007/s10784-008-9085-3.

PMR (Partnership for Market Readiness). 2018. Partnership for Market Readiness. Available at www.thepmr.org/.

R20 (Regions of Climate Action). 2018. About Us. Available at https://regions20.org/about-us-2/#r20-mission.

RE100. 2018. RE100. Available at there100.org/re100.

REEEP (Renewable Energy and Energy Efficiency Partnership). 2018. About REEEP. Available at www.reeep.org/about-reeep.

REN21 (Renewable Energy Policy Network for the 21st Century). 2018. About Us. Available at www.ren21.net/about-ren21/about-us/.

Röhrkasten, S. 2015. *Global Governance on Renewable Energy - Contrasting the Ideas of the German and Brazilian Governments*. Wiesbaden: Springer VS.

RSB (Roundtable on Sustainable Biomaterials). 2018. Certification. 2018. Available at https://rsb.org/certification/.

Sanderink, L., Kristensen, K., Widerberg, O., and Pattberg, P. 2018. *Mapping the Institutional Architecture of Global Energy Governance*. Amsterdam. https://doi.org/10.13140/RG.2.2.22530.38085.

SEforALL. 2018. SEforALL About Us. Available at www.seforall.org/about-us.

Sovacool, B. K. and Florini, A. 2012. Examining the Complications of Global Energy Governance. *Journal of Energy and Natural Resources Law* 30(3), 235–263. https://doi.org/10.1080/02646811.2012.11435295.

Sterner, T. and Coria, J. 2011. *Policy Instruments for Environmental and Natural Resource Management*. New York: Routledge.

Suding, P. H. and Lempp, P. 2007. The Multifaceted Institutional Landscape and Processes of International Renewable Energy Policy. International Association for Energy Economics Newsletter Second Qua: 4–9. Available at www.iaee.org/documents/newsletterarticles/Suding.pdf.

Tol, R. S. J. 2011. The Social Cost of Carbon. *Annual Review of Resource Economics* 3, 419–443. https://doi.org/10.1146/annurev-resource-083110-120028.

United Nations. 2015. General Assembly Resolution 70/1. Transforming Our World: The 2030 Agenda for Sustainable Development.

Van de Graaf, T. 2013. *The Politics and Institutions of Global Energy Governance*. Hampshire & New York: Palgrave Macmillan. https://doi.org/10.1057/9781137320735.

Van de Graaf, T. and Colgan, J. 2016. *Global Energy Governance: A Review and Research Agenda*. Palgrave Communications. https://doi.org/10.1057/palcomms.2015.47.

VERRA. 2018. Standards for a Sustainable Future. Available at https://verra.org/.

WBCSD (World Business Council for Sustainable Development). 2018. Climate & Energy. Available at www.wbcsd.org/Programs/Climate-and-Energy.

WCI (Western Climate Initiative). 2018. Western Climate Initiative. Available at www.wci-inc.org/.

Widerberg, O., Pattberg, P., and Kristensen, K. 2016. *Mapping the Institutional Architecture of Global Climate Change Governance V.2*. Amsterdam.

Wilson, J. D. 2015. Multilateral Organisations and the Limits to International Energy Cooperation. *New Political Economy* 20(1), 85–106. https://doi.org/10.1080/13563467.2013.872611.

3.8 Annex I:

Database of Institutions in the Climate-Energy Nexus

ZONE	ACRONYM	NAME	DATE	MEM-BERS	ACTORS	TYPE	ROLE	THEME
1	ACE	Association of Southeast Asian Nations Centre for Energy	1999	10	Public	Public	3	Multiple themes
1	AEEP	Africa-EU Energy Partnership	2007	6	Public	Public	6	Energy access
1	APEC_EWG	Asia Pacific Economic Cooperation Energy Working Group	1990	21	Public	Public	3	Multiple themes
1	AREI	Africa Renewable Energy Initiative	2015	6	Public	Public	10	Renewable energy
1	BASREC	Baltic Sea Region Energy Cooperation	1998	11	Public	Public	2	Multiple themes
1	CCREEE	Caribbean Center for Renewable Energy and Energy Efficiency	2015	19	Public	Public	6	Multiple themes
1	CEM	Clean Energy Ministerial	2009	24	Public	Public	6	Renewable energy
1	CESC	Clean Energy Solutions Center	2009	1	Public	Public	3	Clean technology
1	CIF	Climate Investment Funds	2008	36	Public	Public	4	Climate mitigation finance
1	CLIMATE ALLIANCE	Climate Alliance of European Cities with Indigenous Rainforest Peoples	1990	1713	Public	Public	1	Energy efficiency
1	CNCA	Carbon Neutral Cities Alliance	2014	20	Public	Public	3	Multiple themes
1	COM	Covenant of Mayors	2008	6115	Public	Public	1	Multiple themes
1	CORSIA	Carbon Offsetting and Reduction Scheme for International Aviation	2016	1	Public	Public	1	Carbon pricing and trading

1	ECO	ECO Partnerships Clean Energy and Energy Efficiency	2008	2	Public	Public	3	Multiple themes
1	ECREEE	Economic Community of West African States Center for Renewable Energy and Energy Efficiency	2008	15	Public	Public	6	Multiple themes
1	EN_CITIES	Energy Cities	1990	179	Public	Public	6	Multiple themes
1	ENERGIA	International Network on Gender and Sustainable Energy	1996	22	Public	Public	2	Energy access
1	ENR	European Energy Network	1991	1	Public	Public	3	Multiple themes
1	EUEI	European Union Energy Initiative	2002	1	Public	Public	3	Multiple themes
1	EUROCITIES	EUROCITIES	2008	99	Public	Public	1	Energy efficiency
1	FFFSR	Friends of Fossil Fuel Subsidy Reform	2010	9	Public	Public	3	Subsidy reform
1	G20_SR	Group of Twenty Subsidy Reform	2009	20	Public	Public	3	Subsidy reform
1	GEEREF	Global Energy Efficiency and Renewable Energy Fund	2008	1	Public	Public	4	Climate mitigation finance
1	GFAAF	Global Framework for Aviation Alternative Fuels	2009	1	Public	Public	1	Transport
1	GMI	Global Methane Initiative	2010	18	Public	Public	6	Renewable energy
1	GNESD	Global Network on Energy for Sustainable Development	2002	2	Public	Public	6	Energy access
1	ICAP	International Carbon Action Partnership	2007	31	Public	Public	3	Carbon pricing & trading

(cont.)

ZONE	ACRONYM	NAME	DATE	MEM-BERS	ACTORS	TYPE	ROLE	THEME
1	ICLEI	Local Governments for Sustainability	1990	1156	Public	Public	8	Multiple themes
1	IEA	International Energy Agency	1974	29	Public	Public	6	Multiple themes
1	IPEEC	International Partnership for Energy Efficiency Cooperation	2009	16	Public	Public	3	Energy efficiency
1	IRENA	International Renewable Energy Agency	2009	149	Public	Public	3	Renewable energy
1	ISCI	International Solar Cities Initiative	2003	5	Public	Public	3	Renewable energy
1	ISGAN	Implementing Agreement for a Co-operative Program on Smart Grids	2010	25	Public	Public	6	Clean technology
1	KP	Kyoto Protocol	1997	192	Public	Public	1	Multiple themes
1	MEF	Major Economies Forum	2009	17	Public	Public	3	Renewable energy
1	MI	Mission Innovation	2015	23	Public	Public	7	Clean technology
1	NEG_ECP	New England Governors and Eastern Canadian Premiers' Annual Conference	1973	11	Public	Public	8	Multiple themes
1	OLADE	Latin American Energy Organization	1973	26	Public	Public	6	Multiple themes
1	PMR	Partnership for Market Readiness	2010	31	Public	Public	7	Carbon pricing & trading
1	R20	Regions of Climate Action	2010	48	Public	Public	6	Renewable energy
1	RECP	Africa-EU Renewable Energy Cooperation Program	2010	2	Public	Public	7	Renewable energy

1	SEAD	Super-Efficient Equipment and Appliance Deployment Initiative	2010	17	Public	Public	6	Clean technology
1	SEADS	European Union Energy Initiative Strategic Energy Advisory and Dialogue Service	2004	1	Public	Public	3	Multiple themes
1	SIDS_DOCK	Small Island Developing States Sustainable Energy and Climate Resilience Initiative	2009	39	Public	Public	7	Multiple themes
1	UN_EN	United Nations Energy	2004	6	Public	Public	1	Energy access
1	UNFCCC	United Nations Framework Convention on Climate Change	1992	195	Public	Public	1	Multiple themes
1	WCI	Western Climate Initiative	2007	4	Public	Public	2	Carbon pricing & trading
1	ZEV	International Zero-Emission Vehicle Alliance	2015	14	Public	Public	8	Transport
2	ACA	Airport Carbon Accreditation	2009	1	Firm	Private	1	Transport
2	BTEC	Break Through Energy Coalition	2015	29	Firm	Private	4	Carbon pricing & trading
2	CATF	Climate Action Takes Flight	2009	1	Firm	Private	1	Transport
2	CNP	CarbonNeutral Protocol	1997	1	Firm	Private	1	Carbon pricing & trading
2	FFC	Fleets for Change	2010	2	Firm	Private	8	Transport
2	GFA	Green Freight Asia	2011	40	Firm	Private	8	Transport
2	GFE	Green Freight Europe	2012	72	Firm	Private	8	Transport
2	GSA	Global Solar Alliance	2015	3	Firm	Private	3	Renewable energy

(cont.)

ZONE	ACRONYM	NAME	DATE	MEM-BERS	ACTORS	TYPE	ROLE	THEME
2	GSEP	Global Sustainability Electricity Partnership (formerly the E8)	1992	11	Firm	Private	6	Renewable energy
2	IATA_COP	International Air Transport Association Carbon Offset Program	2009	1	Firm	Private	1	Carbon pricing & trading
2	IEEA	Industrial Energy Efficiency Accelerator	2001	1	Firm	Private	3	Carbon pricing & trading
2	IETA	International Emissions Trading Association	1999	135	Firm	Private	3	Carbon pricing & trading
2	REC	Renovate Europe Campaign	2011	18	Firm	Private	3	Energy efficiency
2	TCC	Think Climate Coalition	2015	10	Firm	Private	3	Transport
2	VCS	Verified Carbon Standard (formerly the Voluntary Carbon Standard)	2007	1	Firm	Private	1	Carbon pricing & trading
2	VER+	VER+	2007	1	Firm	Private	1	Carbon pricing & trading
2	WBCSD_E&C	World Business Council for Sustainable Development Energy and Climate	1992	1	Firm	Private	6	Multiple themes
3	CLASP	The Collaborative Labelling and Appliance Standards Program	1999	1	CSO	Private	1	Energy efficiency
3	EUROSOLAR	The European Association for Renewable Energy	1988	1	CSO	Private	6	Renewable energy
3	GFF	Go Fossil Free	2014	1	CSO	Private	3	Climate mitigation finance

3	GS	The Gold Standard	2004	1	CSO	Private	1	Multiple themes
3	GSI	Global Subsidies Initiative	2005	1	CSO	Private	6	Subsidy reform
3	INFORSE	International Network for Sustainable Energy	1992	1	CSO	Private	3	Renewable energy
3	WGBC	World Green Building Council	2002	30	CSO	Private	6	Energy efficiency
4	C4C	United Nations Global Compact Caring for Climate	2007	3	Public/ Firm	Hybrid	1	Carbon pricing & trading
4	CPLC	Carbon Pricing Leadership Coalition	2015	138	Public/ Firm	Hybrid	3	Carbon pricing & trading
4	CTIPFAN	Climate Technology Initiative of the Private Financing Advisory Network	2006	7	Public/ Firm	Hybrid	6	Clean technology
4	ENTSOE	European Network of Transmission System Operators	2008	40	Public/ Firm	Hybrid	6	Clean technology
4	GEEAP	Global Energy Efficiency Accelerator Platform	2012	1	Public/ Firm	Hybrid	3	Energy efficiency
4	GGFRP	Global Gas Flaring Reduction Partnership	2002	34	Public/ Firm	Hybrid	8	Energy efficiency
4	LCTPI	Low Carbon Technology Partnerships Initiative	2014	3	Public/ Firm	Hybrid	6	Clean technology
4	SBCI	United Nations Environment Program Sustainable Buildings and Climate Initiative	2010	9	Public/ Firm	Hybrid	6	Energy efficiency
5	GFEI	Global Fuel Economy Initiative	2009	6	Public/ CSO	Hybrid	3	Transport
5	GGFAP	Global Green Freight Action Plan	2015	9	Public/ CSO	Hybrid	3	Transport

(cont.)

ZONE	ACRONYM	NAME	DATE	MEM-BERS	ACTORS	TYPE	ROLE	THEME
6	GSC	Global Solar Council	2015	39	CSO/ Firm	Private	6	Renewable energy
6	ISES	International Solar Energy Society	1954	1	CSO/ Firm	Private	3	Renewable energy
6	PPMC	Paris Process on Mobility and Climate	2015	2	CSO/ Firm	Private	3	Transport
6	RE100	100% Renewables	2014	2	CSO/ Firm	Private	1	Renewable energy
6	RN!	Refrigerants, Naturally!	2004	5	CSO/ Firm	Private	3	Clean technology
6	SE	Shipping Efficiency	2001	2	CSO/ Firm	Private	3	Transport
7	CAA	Clean Air Asia	2001	251	Public/ CSO/ Firm	Hybrid	3	Transport
7	CUD	Connected Urban Development	2006	2	Public/ CSO/ Firm	Hybrid	3	Energy efficiency
7	DI	Divest Invest Global Movement	2014	41	Public/ CSO/ Firm	Hybrid	7	Climate mitigation finance
7	E4I	Energy for Impact (formerly GVEP)	2005	19	Public/ CSO/ Firm	Hybrid	6	Energy access
7	ENERGY+	International Energy and Climate Initiative	2010	43	Public/ CSO/ Firm	Hybrid	4	Energy access
7	ENLIGHT	En.Lighten	2009	25	Public/ CSO/ Firm	Hybrid	6	Clean technology
7	GACC	Global Alliance for Clean Cookstove	2010	1615	Public/ CSO/ Firm	Hybrid	6	Energy access
7	GBEP	Global Bioenergy Partnership	2007	37	Public/ CSO/ Firm	Hybrid	6	Renewable energy
7	L&G	Lean and Green	2008	1	Public/ CSO/ Firm	Hybrid	8	Transport
7	LEDS_GP	Low Emission Development Strategies Global Partnership	2011	27	Public/ CSO/ Firm	Hybrid	3	Energy access

7	NCM	Networked Carbon Markets Initiative	2013	1	Public/ CSO/ Firm	Hybrid	3	Carbon pricing & trading
7	PCFV	United Nations Environment Program Partnership for Clean Fuels and Vehicles	2002	76	Public/ CSO/ Firm	Hybrid	3	Transport
7	PDC	Portfolio Decarbonization Coalition	2014	4	Public/ CSO/ Firm	Hybrid	6	Climate mitigation finance
7	REEEP	Renewable Energy and Energy Efficiency Partnership	2002	354	Public/ CSO/ Firm	Hybrid	4	Multiple themes
7	REN21	The Renewable Energy Policy Network for the 21st Century	2005	52	Public/ CSO/ Firm	Hybrid	3	Renewable energy
7	RSB	The Roundtable on Sustainable Biofuels (RSB Standard)	2007	80	Public/ CSO/ Firm	Hybrid	1	Renewable energy
7	SEFORALL	Sustainable Energy for All	2011	2	Public/ CSO/ Firm	Hybrid	1	Energy access
7	SLOCAT	Partnership on Sustainable Low Carbon Transport	2009	94	Public/ CSO/ Firm	Hybrid	3	Transport
7	U4E	United for Efficiency	2010	18	Public/ CSO/ Firm	Hybrid	3	Energy efficiency
7	UEMI	Urban Electric Mobility Initiative	2014	21	Public/ CSO/ Firm	Hybrid	3	Transport

3.9 Annex II:

Descriptions of Institutions in the Climate-Energy Nexus

ACRONYM	DESCRIPTION
ACA	Airport Carbon Accreditation is an independent programme to enforce the accreditation criteria for airports on an annual basis. The ACA aims to reduce carbon emissions and to increase airport sustainability.
ACE	The ASEAN Centre for Energy serves as a high-performing institution, a regional centre of excellence that builds a coherent, coordinated, focused, and robust energy policy agenda and strategy for the Association of Southeast Asian Nations Centre for Energy, with three roles: an ASEAN Energy Think Tank; a catalyst to unify and strengthen ASEAN Energy Cooperation and Integration; and an ASEAN Energy Data and Knowledge Hub.
AEEP	Established in 2007 as one of the partnerships under the Joint Africa-EU Strategy, the Africa-EU Energy Partnership (AEEP) is a long-term framework for strategic dialogue between Africa and the EU aimed at sharing knowledge, setting political priorities and developing joint programmes on the key energy issues and challenges in the twenty-first century.
APEC_EWG	The work of the APEC Energy Working Group (EWG) aims to strengthen energy security; promote energy efficiency and sustainable communities; develop cleaner energy source; and enhance trade and investment in all energy sources to promote economic prosperity.
AREI	The Africa Renewable Energy Initiative (AREI) aims at enabling the installation of large-scale renewable energy capacity on the African continent by 2020, which would have a considerable impact on the reduction of greenhouse gas emissions in the continent.
BASREC	BASREC represents a unique and important regional forum for dialogue on energy policy and global climate change issues with an emphasis on the promotion of energy efficiency, the use of renewable energy and other sustainable supply sources.
BTEC	The Breakthrough Energy Coalition is a global group of twenty-eight high net worth investors committed to funding clean energy companies that are emerging from the initiatives of Mission Innovation, which was announced at the 2015 COP21. The group aims to bolster governmental assistance in renewable energy to 20 billion US dollars.

(*cont.*)

ACRONYM	DESCRIPTION
C4C	UN Global Compact, UNEP, and the secretariat of the UNFCCC's initiative to advance the role of business in addressing climate change. It provides a framework for business leaders to implement practical climate change solutions and help shape public policy.
CAA	Clean Air Asia is an international nongovernmental organization that leads the regional mission for better air quality and healthier, more liveable cities in Asia. It aims to reduce air pollution and greenhouse gas emissions in 1000+ cities in Asia through policies and programmes that cover air quality, transport, and industrial emissions and energy use. It works with ministries (energy, environment, health and transport), cities in Asia, the private sector, and development agencies to provide leadership and technical knowledge in the following areas: Air Quality and Climate Change, Low Emissions Urban Development, Clean Fuels and Vehicles, and Green Freight and Logistics.
CATF	In 2008, the aviation industry presented the world's first global transport sector climate action framework, based on a set of three global goals, underpinned by four pillars of climate action. The framework set out aims at 1.5 per cent average annual fuel efficiency improvement from 2009 to 2020; stabilizing net aviation CO_2 emissions at 2020 levels through carbon neutral growth; and reducing aviation's net CO_2 emissions to 50 per cent of what they were in 2005 in 2050.
CCREEE	CCREEE aims at improving access to modern, affordable, and reliable energy services, energy security, and mitigation of negative externalities of the energy system (e.g. local pollution and GHG emissions) by promoting renewable energy and energy-efficiency investments, markets, and industries in the Caribbean. The centre complements and strengthens ongoing national/regional activities in the areas of policy and capacity development, knowledge management, and awareness rising, as well as investment and business promotion.
CEM	Global forum to share best practices and promote policies and programmes that encourage and facilitate the transition to a global clean energy economy. Its initiatives help reduce emissions, improve energy security, provide energy access, and sustain economic growth.
CESC	The Clean Energy Solutions Center helps governments, advisors, and analysts create policies and programmes that advance the deployment of clean energy technologies. The Solutions Center is an initiative of the Clean Energy Ministerial (CEM), a global forum to share best practices and promote policies and programmes that encourage and facilitate the transition to a global clean energy economy.
CIF	Provides developing and middle-income countries with urgently needed resources to mitigate and manage the challenges of climate change and reduce their greenhouse gas emissions; since 2008, it champions innovative country-led investments in clean technology, renewable energy, sustainable management of forests, and climate-resilient development.

(*cont.*)

ACRONYM	DESCRIPTION
CLASP	CLASP works hand-in-hand with policy makers, governments, technical experts, industry, funding organizations, consumers and consumer groups, and others to improve the environmental and energy performance of the appliances and related systems we use every day, lessening their impacts on people and the world around us.
CLIMATE_ ALLIANCE	Association of cities, municipalities, and districts committed to the protection of the global climate, aiming to reduce greenhouse emissions. For this, local climate strategies are developed and implemented, especially in the energy and transport sectors.
CNCA	The Carbon Neutral Cities Alliance (CNCA or 'Alliance') aims to address what it will take for leading international cities to achieve these deep emissions reductions and how they can work together to meet their respective goals more efficiently and effectively.
CNP	Market leaders and pioneers in the world of carbon neutral certification and carbon reduction; it provides a robust framework and credible certification that a company, brand, or product has reduced their carbon emissions to net zero.
COM	European network involving local and regional authorities, voluntarily committing to increasing energy efficiency and use of renewable energy sources on their territories. By their commitment, Covenant signatories aim to meet and exceed the European Union 20 per cent CO_2 reduction objective by 2020.
CORSIA	Under the Carbon Offsetting Scheme for International Aviation (CORSIA), aircraft operators will be required to purchase offsets, or 'emission units', for the growth in CO_2 emissions covered by the scheme. CORSIA aims to address any annual increase in total CO_2 emissions from international civil aviation above 2020 levels.
CPLC	The Coalition is a voluntary partnership of national and sub-national governments, businesses, and civil society organizations that agree to advance the carbon pricing agenda by working with each other toward the long-term objective of a carbon price applied throughout the global economy. The Coalition will collect the evidence base, benefiting from experience around the world in designing and using carbon pricing, and use this input to help inform successful carbon pricing policy development and use of carbon pricing in businesses.
CTIPFAN	Multilateral public–private partnership, initiated by the Climate Technology Initiative and the UNFCCC, which connects clean energy businesses and projects with private sector financing. Through its network of private sector consultants, it provides targeted professional support and advice and technical assistance to selected projects on the preparation of commercially viable, sustainable, and climate-friendly business models for introduction to investors.
CUD	CUD demonstrates how to reduce carbon emissions by introducing fundamental improvements in the efficiency of urban infrastructure through information and communications technology. It was born from Cisco's commitment to the Clinton Global Initiative to participate in helping reduce carbon emissions.

(*cont.*)

ACRONYM	DESCRIPTION
DI	D-I encourages inventors across the public and private sectors and across different types of funds to divest from fossil fuel industries and promote a clean energy future. In 2015 D-I represented more than 500 organizations and US $3.4 trillion.
E4I	E4I (formerly GVEP) believes in a private sector approach to development, utilizing donor funding to help businesses succeed and grow. E4I focuses on supporting the development of business models to deliver energy access the enterprise approach, and believes it will result in long-lasting change and sustainable results.
ECO	The mission of the EcoPartnerships initiative is to elevate successful sub-national cooperation models to international prominence, and by doing so, to spur broad replication by their peers in the U.S. and China. The U.S.–China EcoPartnerships programme offers sub-national organizations from each country a unique opportunity to pair up and demonstrate breakthrough clean energy, climate change, and environmental solutions.
ECREEE	ECREEE aspires to contribute to the sustainable economic, social, and environmental development of West Africa by improving access to modern, reliable, and affordable energy services, energy security, and reduction of negative environmental externalities of the energy system. ECREEE aims to create favourable framework conditions and an enabling environment for renewable energy and energy efficiency markets by supporting activities directed at mitigating existing barriers within the technological, financial, economic, business, legal, policy, institutional, knowledge, and capacity-building framework.
EN_CITIES	European Association of local authorities in energy transition; its objectives are: to strengthen society's role and skills in the field of sustainable energy, to represent people's interests and influence the policies and proposals made by EU institutions in the field of energy, environmental protection, and urban policy, and to develop and promote people's initiatives through exchange of experiences, the transfer of know-how, and the implementation of joint projects.
ENERGIA	ENERGIA believes that projects, programmes, and policies that explicitly address gender and energy issues have better outcomes and improve the livelihood of entire communities. By involving women in the development delivery and use of modern energy, sustainability and adoption rates of these services are enhanced. In order to provide continued support and have gender be part of the developmental process, ENERGIA also creates unique training modules and tools for the energy sector.
ENERGY+	Led by the Norwegian Ministry of Foreign Affairs, the overarching purpose of the Energy+ Initiative is to contribute to providing access to efficient energy services to all by increased development of renewable energy and energy efficiency, and to mitigate energy's impacts on climate. It is an open partnership engaging, in particular countries in the developing world.

(*cont.*)

ACRONYM	DESCRIPTION
ENLIGHT	Initiative by UNEP and GEF to accelerate a global market transformation to environmentally sustainable, energy-efficient lighting technologies, as well as to develop strategies to phase out inefficient incandescent lamps to reduce CO_2 emissions and the release of mercury from fossil fuel combustion. It serves as a platform to build synergies among international stakeholders; identify global best practices and share this knowledge and information; create policy and regulatory frameworks; address technical and quality issues; and encourage countries to develop National and/or Regional Efficient Lighting Strategies.
ENR	EnR is a voluntary network with responsibility for the planning, management, or review of national research, development, demonstration, or dissemination programmes in the fields of energy efficiency and renewable energy and climate change abatement. It provides a first point of contact for national energy agencies in EU Member States. EnR dedicates its efforts toward joint activities where its unique character provides added value at both a European and individual Member State level.
ENTSOE	As the legally mandated body of electricity TSOs at the European level, ENTSO-E's mission is to fulfil its various legal mandates for the benefit of electricity customers and to leverage its mandated work products to shape future energy policy for the benefit of society. It aims to facilitate secure integration of new generation sources, particularly renewable energy, as well as significantly contributing to the EU's greenhouse gases reduction and renewable energy supply goals.
EUEI	The EUEI PDF is a multi-donor facility that contributes to the achievement of the Sustainable Development Goals, in particular on energy. As a flexible instrument of the European Union, EUEI PDF promotes sustainable energy for equitable development in Africa, Latin America, and Asia. Therefore, it facilitates energy dialogue and knowledge transfer; advises partners to create enabling environments for sustainable energy solutions; supports the development of sustainable energy markets; and conducts and promotes research, innovation, and capacity development.
EUROCITIES	Network of major European cities; they offer members a platform for sharing knowledge and exchanging ideas. They influence and work with EU institutions to respond to common issues that affect the day-to-day lives of Europeans. Their goal is to reinforce the important role that local governments should play in a multilevel governance structure.
EUROSOLAR	EUROSOLAR conducts its work independently of political parties, institutions, commercial enterprises, and interest groups and is a registered non-profit organization. It is dedicated to the cause of completely substituting for nuclear and fossil energy through renewable energy. EUROSOLAR acts bring together expertise and

(cont.)

ACRONYM	DESCRIPTION
	develops and encourages political and economic action plans. Additionally, it addresses and initiates action at the international level, and also the national, regional, and local level.
FFC	Fleets for Change works with the transportation sector to reducing GHGs in North America through implementing measures to increase fuel efficiency, reduce mileage, switch to low-carbon fuels, and to use new technology vehicles.
FFFSR	The FFFSR is in informal group of non-G20 countries working to build political consensus on the importance of fossil fuel subsidy reform. The Friends work internationally within forums such as the G20, APEC, OECD, World Bank, UNFCCC and the UN Sustainable Development Agenda to convince governments of the benefits of reform, and to help them with ways to do it. The friends advocate that reform needs to be ambitious and transparent.
G20_SR	The G20 Subsidy Reform aims to remove fossil fuel subsidies internationally, by means of a general political dialogue.
GACC	The Global Alliance for Clean Cook stoves (Alliance), hosted by the UN Foundation, is at the forefront of efforts to promote the adoption of clean cooking solutions and spur universal adoption of clean cook stoves and fuels. The Alliance and its partners are working to establish a thriving global market for clean cooking solutions by addressing the market barriers that impede the production, deployment, and use of clean and efficient cook stoves, and fuels in developing countries.
GBEP	Partnership that brings together public, private, and civil society stakeholders in a joint commitment to promote bioenergy for sustainable development. It focuses its activities in three strategic areas: sustainable development, climate change, and food and energy security.
GEEAP	Sustainable Energy for All is an initiative led by the UN secretary-general and the president of the World Bank, has as one of its three objectives for 2030 a doubling of the global rate of improvement in energy efficiency. The Global Energy Efficiency Accelerator Platform was established to help reach this objective. It will do so by driving action and commitments by national and sub-national leaders at the country, city, state, region, or sector level. A key deliverable will be Integrated Policy and Investment Roadmaps prepared with committed public and private partners. These Roadmaps will guide project implementation supported by a global network of experts, institutions, and businesses.
GEEREF	Advised by the European Investment Bank Group, GEEREF is an innovative Fund-of-Funds catalyzing private sector capital into clean energy projects in developing countries and economies in transition.
GFA	GFA engages companies using road freight services and companies that own commercial road freight fleets. The key objective of the institution is to lower GHG emissions through decreasing fuel

(*cont.*)

ACRONYM	DESCRIPTION
	consumption. It does so by sharing information on best practices regarding green technology, by providing a level playing field for companies to reduce emissions from their transport segment, and by providing a clear definition of what green transport is.
GFAAF	GAAF promotes the use of sustainable aviation alternative fuels is a key part of the basket of measures under consideration by ICAO Member States to achieve the aspirational goal of stabilizing emissions from international aviation at their 2020 levels. ICAO is actively engaged in activities facilitating, on a global basis, the promotion and harmonization of initiatives that encourage and support the development of sustainable alternative fuels for international aviation.
GFE	Similar to GFAN, GFE is an industry-led programme including companies engaging with or engaged in transportation over land and sea. It encourages GHG emission reduction by establishing an emissions monitoring and reporting platform, by promoting collaboration between carriers and shippers, and finally by encouraging engagement through certification.
GFEI	The Global Fuel Economy Initiative (GFEI) assist governments and transport stakeholders promote greater fuel economy. Using the skills and expertise of the GFEI partners, the GFEI Toolkit team are able to establish a baseline in each country; present policy options and case studies; and enable all stakeholders to engage in the policy process.
GFF	GFF is an international network of campaigns and campaigners working toward freeing communities from fossil fuels. While each campaign is independently run and may bring different emphases depending on their local context, the majority of campaigns are asking institutions to: immediately freeze any new investment in fossil fuel companies; divest from direct ownership and any commingled funds that include fossil fuel public equities and corporate bonds within five years; and end their fossil fuels sponsorship.
GGFAP	The goal of the Global Green Freight Action Plan is to enhance the environmental energy efficiency of goods movement in ways that significantly reduce the climate, health, energy, and cost impacts of freight transport around the world. Full implementation of the Action Plan will shape a more sustainable global freight sector where goods, materials, and trade flows move with the best available technologies and strategies through an efficient, cleaner and greener, multimodal, global freight supply chain. Performance data and best practices will be shared and exchanged via green freight programmes and in ways that enhance efficiency, cost savings, competitiveness, environmental performance, public health, and economic development.
GGFRP	The Global Gas Flaring Reduction Partnership (GGFR) is a public–private initiative comprising international and national oil companies, national and regional governments, and international institutions. GGFR works to increase use of natural gas associated with oil production by helping remove technical and regulatory barriers

(cont.)

ACRONYM	DESCRIPTION
	to flaring reduction, conducting research, disseminating best practices, and developing country-specific gas flaring reduction programmes.
GMI	The GMI is an international public–private initiative that advances cost-effective, near-term methane abatement and recovery projects and the use of methane as a clean energy source. Activities of the GMI are focused on reducing informational, institutional, and market barriers to project development by making available tools and resources, providing training and capacity building, conducting technology demonstrations, and offering direct project support.
GNESD	The Global Network on Energy for Sustainable Development (GNESD) is a UNEP facilitated knowledge network of Member Centres and network partners worldwide, renowned for their work on energy, development, and environment issues. Member Centers and Associates coordinate joint activities within these fields, exchange information, carry out analytical studies, and supply policy support.
GS	Gold Standard is a voluntary carbon offsetting standard. To receive the stamp of approval, all Gold Standard projects must be implemented following best practice rules, consult with local stakeholders, continually reduce greenhouse gas emissions, and improve the environment and people's lives. Once certified, their projects are issued credits annually against independently audited climate and sustainable development outcomes. The purchase of these credits – by governments, business, impact investors, and individuals – provides on-going funding to project activities.
GSA	The GSA is a community-oriented project initiated by leading trade exhibitions & business conferences in Europe, China, USA, with the purpose of driving the global development of the solar markets and industry. It strives to spread awareness, information, and advocacy among the professional community, the decision makers and the general public while promoting solar energy as a mainstream solution for a low-carbon economy.
GSC	The Global Solar Council was established by leading regional and national solar associations. It will unify the entire solar power sector at an international level, share best practices, and work collaboratively to accelerate solar electricity deployment worldwide.
GSEP	GSEP is a not-for-profit organization whose members are the world's leading electricity companies. It promotes sustainable energy development through electricity sector projects and human capacity building activities in developing and emerging nations worldwide.
GSI	Established in 2005 by the International Institute for Sustainable Development (IISD), the Global Subsidies Initiative (GSI) is dedicated to analyzing subsidies – transfers of public money to private interests – and how they support or undermine efforts to achieve sustainable development.
IATA_COP	IATA encourages the use of voluntary initiatives to address environmental impacts from aviation and promotes the use of industry

(*cont.*)

ACRONYM	DESCRIPTION
	best practices where possible. Voluntary initiatives can be tailored to the specific needs of governments, industry, and other stakeholders. They can provide more flexibility and cost savings than regulatory measures. In order to further reduce aviation's impact on the environment, IATA has set up several programmes to assist airlines in improving their environmental performance.
ICAP	ICAP is an international forum for governments and public authorities that have implemented or are planning to implement emissions trading systems. It facilitates cooperation and best practices-sharing between countries, sub-national jurisdictions, and supranational institutions that have established or are actively pursuing carbon markets through mandatory cap and trade systems.
ICLEI	World's leading network of more than 1,000 cities, towns, and metropolises committed to building a sustainable future. By helping Members to make their cities and regions sustainable, low-carbon, resilient, eco-mobile, biodiverse, resource-efficient and productive, healthy and happy, with a green economy and smart infrastructure, they impact more than 20 per cent of the world's urban population.
IEA	The IEA is an autonomous organization that works to ensure reliable, affordable, and clean energy for its twenty-nine member countries and beyond. The IEA has four main areas of focus: energy security, economic development, environmental awareness, and engagement worldwide.
IEEA	The Industrial Energy Efficiency Accelerator (IEEA) works collaboratively with trade associations, sector companies, and technology providers to identify innovative opportunities for energy reduction. Working with key stakeholders in each sector they have identified innovations in equipment, processes, and product strategy. Their work with fourteen mid-energy intense industry sectors has identified energy, carbon emissions reduction averaging 29 per cent.
IETA	IETA is a non-profit business organization created to establish a functional international framework for trading in greenhouse gas emission reductions. Membership includes leading international companies from across the carbon trading cycle; they seek to develop an emissions trading regime that results in real and verifiable greenhouse gas emission reductions, while balancing economic efficiency with environmental integrity and social equity.
INFORSE	INFORSE is a global network of independent nongovernmental organizations working for sustainable energy solutions to reduce poverty and protect the environment. The aim of INFORSE is to raise awareness and provide advocacy; to build up capacity at local, national, and international level; to work for institutional reform; and to support research and development.
IPEEC	Autonomous international forum that provides global leadership on energy efficiency by facilitating government implementation of policies and programmes to yield energy-efficient gains. It is

(cont.)

ACRONYM	DESCRIPTION
	dedicated to facilitating rapid deployment of clean energy technologies worldwide and promoting information exchange on best practices to facilitate initiatives that improve energy efficiency. It has been identified as the lead coordinating organization to carry out the G20 Energy Efficiency Action Plan.
IRENA	IRENA is an Intergovernmental organization that supports countries in their transition to a sustainable energy future and serves as the principal platform for international cooperation, a centre of excellence, and a repository of policy, technology, resource, and financial knowledge on renewable energy. It promotes the widespread adoption and sustainable use of all forms of renewable energy, including bioenergy, geothermal, hydropower, ocean, solar, and wind energy in the pursuit of sustainable development, energy access, energy security, low-carbon economic growth, and prosperity.
ISCI	The International Solar Cities Initiative (ISCI) is an international non-profit organization dedicated to promoting new urban policies, planning, and practices that reduce city per capita greenhouse gas emissions to levels consistent with long term climate sustainability as estimated by the IPCC.
ISES	The underlying goal behind the work of ISES is to advance the transition to a renewable energy world. ISES is committed to 100 per cent renewable energy for all used efficiently and wisely. ISES provides key timely information on renewable energy technology and innovation breakthroughs, policy mechanisms and changes, investment strategies and deployment opportunities.
ISGAN	ISGAN creates a mechanism for multilateral government-to-government collaboration to advance the development and deployment of smarter electric grid technologies, practices, and systems. It aims to improve the understanding of smart-grid technologies, practices, and systems, and to promote adoption of related enabling government policies. ISGAN facilitates dynamic knowledge sharing, technical assistance, and project coordination, where appropriate.
KP	International agreement linked with the UNFCCC, which commits its Parties by setting internationally binding emission-reduction targets.
L&G	L&G has developed a simulation that encourages companies and government bodies to reduce their carbon footprint through taking cost saving measures. Furthermore, the institution provides awards to companies that prove they can reduce emission by 20 per cent over five years, and a star upon completion.
LCTPI	With a solid framework and clear agenda, LCTPi is a unique, action-oriented programme that brings together companies and partners to accelerate the development of low-carbon technology solutions to stay below the 2°C ceiling. LCTPi has gathered more than 150 global businesses with 70 partners to work collaboratively on the climate challenge.
LEDS_GP	LEDS GP aims at reducing GHG emissions while also increasing resilience toward climate change impacts. It does so by linking

(*cont.*)

ACRONYM	DESCRIPTION
	practitioners and policy makers in regional platforms and work groups that promote low emission development strategies. It has six workgroups working on topics including Agriculture and Forestry, Energy, Finance, and Transportation.
MEF	17 major economies forum; it is intended to facilitate a candid dialogue among major developed and developing economies, help generate the political leadership necessary to achieve a successful outcome at the annual UN climate negotiations, and advance the exploration of concrete initiatives and joint ventures that increase the supply of clean energy while cutting greenhouse gas emissions.
MI	Mission Innovation (MI) is a global initiative of twenty-two countries and the European Union to dramatically accelerate global clean energy innovation. As part of the initiative, participating countries have committed to double their governments' clean energy research and development (R&D) investments over five years, while encouraging greater levels of private sector investment in transformative clean energy technologies. These additional resources will dramatically accelerate the availability of the advanced technologies that will define a future global energy mix that is clean, affordable, and reliable.
NCM	Through the Networked Carbon Markets Initiative, the World Bank Group is convening civil society, governments, and the private sector to develop a framework for assessing climate mitigation efforts and infrastructure to support carbon market related functions. The end-goal is to facilitate linking or 'networking' of heterogeneous carbon markets so that the linked markets will have greater liquidity and deliver climate-smart financing more efficiently.
NEG_ECP	Nonpartisan association of the seven governors of Northeast states: Connecticut, Maine, Massachusetts, New Hampshire, New York, Rhode Island, and Vermont. It encourages intergovernmental cooperation on issues affecting the economic, social, and environmental well-being of the Northeast. In the region, it is a forum for states to exchange information and undertake cooperative action on issues of mutual interest.
OLADE	OLADE aims to contribute to the integration, sustainable development, and energy security in the region, advising and promoting cooperation and coordination among its member countries. OLADE is the political and technical-support organization by means of which its Member States undertake common efforts to achieve regional and sub-regional energy integration.
PCFV	The Partnership for Clean Fuels and Vehicles (PCFV) is the leading global public–private initiative promoting cleaner fuels and vehicles in developing and transition countries. Established at the World Summit on Sustainable Development in September 2002 in Johannesburg, the PCFV brings together seventy-two organizations representing developed and developing countries, the fuel and vehicle

(cont.)

ACRONYM	DESCRIPTION
	industries, civil society, and leading world experts on cleaner fuels and vehicles. The partners combine their resources and efforts to achieve cleaner air and lower greenhouse gas emissions from road transport by applying fuel quality improvements and proven vehicle technologies in use in leading global auto markets.
PDC	Portfolio decarbonization can be achieved by withdrawing capital from particularly carbon-intensive companies, projects, and technologies in each sector and by reinvesting that capital into particularly carbon-efficient companies, projects, and technologies of the same sector. It can also be achieved through targeted engagement by investors with portfolio companies. When large institutional investors start to engage and/or reallocate capital on the basis of companies' GHG emissions, it provides a strong incentive for those companies to re-channel their own investments from carbon-intensive to low-carbon activities, assets, and technologies.
PMR	PMR is a Forum for collective innovation and action and fund to support capacity building to scale up climate change mitigation. it provides support to prepare and implement climate change mitigation policies – carbon pricing instruments – in order to scale up GHG mitigation. Serving as a platform to share lessons, countries work together to shape the future of cost-effective GHG mitigation.
PPMC	The PPMC is an open and inclusive platform that actively invites all organizations and initiatives that support effective action on transport and climate change to join in the process. The PPMC was created in early 2015 to strengthen the voice of the sustainable transport community in the UNFCCC process. The PPMC will engage global processes on sustainable development and climate change to ensure that implementation arrangements are conducive for action by the transport sector.
R20	To help sub-national governments around the world to develop low-carbon and climate-resilient economic development projects. It aims to help build an effective green deal flow at sub-national level by connecting Regions, Technology and Finance to build sustainable low-carbon projects.
RE100	RE100 contributes to global GHG mitigation through encouraging its members from the private sector to go shift their electricity supply to 100 per cent renewable energy sources. For companies that are not ready for this commitment RE100 helps overcome barriers and develop transparent reporting schemes.
REC	The Renovate Europe Campaign (REC), launched in 2011, is an initiative of EuroACE, the European Alliance of Companies for Energy Efficiency in Buildings. It is the only EU-wide campaign that focuses exclusively on ambitious renovation of the building stock in the EU and is the voice that 'bangs the drum' for energy-efficient renovations, taking a technology neutral, integrated and holistic approach to energy-efficient renovations.

(*cont.*)

ACRONYM	DESCRIPTION
RECP	The RECP supports market development in a variety of ways. This includes the provision of critical information on African energy markets, identification of concrete project opportunities through on-the-ground scouting activities, matchmaking between project developers, technology suppliers, and service providers for joint project and business development in Africa as well as facilitation of access to finance. In addition to these private sector focused activities, RECP also provides policy advisory services and supports local skills development by working with technical and vocational training institutions and academia.
REEEP	REEEP invests in clean energy markets in developing countries to reduce CO_2 emissions and build prosperity. Based on a strategic portfolio of high impact projects, it works to generate energy access, improve lives and economic opportunities, build sustainable markets, and combat climate change.
REN21	Global renewable energy policy multi-stakeholder network that connects a wide range of key actors from Governments, International organizations, Industry associations, and science and academia as well as civil society, to facilitate knowledge exchange, policy development, and joint action toward a rapid global transition to renewable energy. It promotes renewable energy to meet the needs of both industrialized and developing countries that are driven by climate change, energy security, development, and poverty alleviation.
RN!	Initiative of international companies taking action against global warming and ozone layer depletion. They replace harmful greenhouse gases in our point-of-sales cooling and freezing units with climate-friendly natural refrigerants. The goal is to make them the preferred cooling technology – in a safe, reliable, and cost-effective manner.
RSB	International multi-stakeholder initiative that brings together farmers, companies, nongovernmental organizations, experts, governments, and intergovernmental agencies concerned with ensuring the sustainability of biomass, and biomaterial production and processing. Their certification system is based on sustainability standards encompassing environmental, social, and economic principles and criteria.
SBCI	The United Nation's Environment Programme's Sustainable Building and Climate Initiative (UNEP-SBCI) is a partnership of major public and private sector stakeholders in the building sector, working to promote sustainable building policies and practices worldwide.
SE	ShippingEfficiency.org is an initiative launched by the Carbon War Room and RightShip to increase information flows around the energy efficiency of international shipping and ultimately help reduce the environmental impacts of the world's shipping fleet.
SEAD	SEAD is about governments working together to save energy, turning knowledge into action to advance global market transformation for

(cont.)

ACRONYM	DESCRIPTION
	energy-efficient products. The SEAD initiative support this effort by providing knowledge and tools; raising awareness; identifying and highlighting technologies; and providing technical expertise.
SEADS	The EUEI PDF offers Strategic Energy Advisory and Dialogue Services (SEADS) to support the development and improvement of energy policies, strategies, and regulations in order to create an enabling environment for sustainable energy investments. The EUEI PDF has been a pioneer in supporting policy change to create favourable frameworks for sustainable energy market development in developing countries.
SEFORALL	Global initiative that brings together top-level leadership from all sectors of society – governments, business, and civil society – to mobilize action from all sectors of society in support of three interlinked objectives: providing universal access to modern energy services, doubling the global rate of improvement in energy efficiency, and doubling the share of renewable energy in the global energy mix.
SIDS_DOCK	SIDS DOCK aims to help develop a sustainable energy sector in small islands, providing the foundation for low-carbon economic growth and adaptation to climate change, with the aim of helping small islands achieve by 2033 50 per cent electric power from renewable sources, a 25 per cent decrease in conventional transportation fuel use, and a 25 per cent increase in energy efficiency (using a 2005 baseline).
SLOCAT	Multi-stakeholder partnership of more than eighty organizations (representing UN organizations, Multilateral and Bilateral development organizations, NGOs and Foundations, Academe, and the Business Sector). It is a Type II Partnership under the UN, meaning that it is a non-legal and non-binding partnership, established to provide a global voice on Sustainable Transport.
TCC	Think Climate is a multi-stakeholder coalition of ten associations with interests in waterborne transport infrastructure. By furthering understanding, providing targeted technical support, and building capacity, the coalition's 'Navigating a Changing Climate' initiative will encourage the owners, operators, and users of waterborne transport infrastructure to: reduce greenhouse gas emissions and shift to low carbon maritime and inland navigation infrastructure; and act urgently to strengthen resilience and improve preparedness to adapt to the changing climate.
U4E	U4E contributes to climate governance by encouraging global markets to switch to more energy-efficient lighting, equipment, and appliances. The institution works under the SE4ALL initiative.
UEMI	UEMI aims at phasing out conventional vehicles to be replaced by at least 30 per cent electric vehicles by 2030. Further, UEMI aims to widen the concept of urban sustainability and a 30 per cent GHG emissions reduction in urban areas by 2030. In doing so, UEMI is also

(*cont.*)

ACRONYM	DESCRIPTION
	developing tools to integrate e-mobility into society, to make a 2-degree pathway, and to assess the impact of electric vehicles.
UN_EN	UN-Energy was initiated as a mechanism to promote coherence within the United Nations family of organizations in the energy field and to develop increased collective engagement between the United Nations and other key external stakeholders. Its envisaged role was to increase the sharing of information, encourage and facilitate joint programming, and develop action-oriented approaches to coordination.
UNFCCC	International environmental treaty with the objective of stabilizing greenhouse gas concentrations in the atmosphere at a level that would prevent dangerous anthropogenic interference with the climate system; it provides a framework for negotiating aiming limits GHG emissions.
VCS	World's leading voluntary greenhouse gas programme founded by a collection of business and environmental leaders who saw a need for greater quality assurance in voluntary carbon markets.
VER+	The VER Plus (VER+) is a carbon offset standard and that follows the Kyoto Protocol's project-based mechanisms (CDM and JI). It was developed by TÜV SÜD.
WBCSD_E&C	The Energy and Climate focus area of the WBCSD provides members with a platform to engage with their peers and stakeholders in energy and climate, to address critical industry issues, and to share ways to solutions. The project delivers business input to the design and implementation of the post-Kyoto climate architecture through an active involvement into international processes.
WCI	Collaboration of independent jurisdictions in North America working together to identify, evaluate, and implement emissions trading policies to tackle climate change at a regional level. This is a comprehensive effort to reduce greenhouse gas pollution, spur investment in clean-energy technologies that create green jobs, and reduce dependence on imported oil.
WGBC	WGBC fosters and supports new and emerging Green Building Councils by providing them with the tools and strategies to establish strong organizations and leadership positions in their countries. By driving collaboration and increasing the profile of the green building market, the WGBC works with its member councils to ensure that green buildings are a part of any comprehensive strategy to deliver carbon emission reductions.
ZEV	The International Zero-Emission Vehicle Alliance (ZEV Alliance) is a collaboration of national and sub-national governments working together to accelerate adoption of ZEVs. The participants set ambitious, achievable targets for ZEV deployment, take actions to achieve those targets as appropriate in each jurisdiction, act together to achieve individual and collective targets, and encourage and support other jurisdictions in setting and achieving ambitious ZEV targets.

Part II

Coherence and Management in the Climate-Energy Nexus

4

Renewable Energy

A Loosely Coupled System or a Well-Connected Web of Institutions?

LISA SANDERINK

4.1 Introduction

Energy is the lifeblood of modern society: it is required to fulfil people's basic needs and everyday activities, and, in the same vein, the world's economic processes heavily rely on energy. However, global energy consumption and production is putting high pressure on the earth system and is arguably the main culprit behind climate change; fossil-fuel combustion accounts for two-thirds of total greenhouse gas (GHG) emissions and 80 per cent of carbon dioxide (IEA/OECD 2017). Therefore, decarbonization of global energy systems is of paramount importance for a sustainable future, and a global uptake of renewable energy plays a key role in this trajectory (e.g. Ki-moon 2011; IRENA 2015; WFC 2016).

While the overall share of renewables in total final energy consumption grew to around 19 per cent and reached a new record in 2017, this growth must accelerate to reach a two-thirds share by 2050 (IRENA 2018a). This is both technically and economically feasible, yet it requires effective global governance to get governments committed, to put regulatory frameworks in place, and to facilitate knowledge exchange and technology transfer (Röhrkasten 2015). As discussed in Chapter 3, the renewable energy subfield is institutionally complex. It is governed by a wide range of different institutions, including international organizations, alongside private institutions and multi-stakeholder partnerships. On top of that, the subfield covers different renewable energy sources, such as solar and wind, and has to navigate three critical challenges, commonly known as energy security, energy access, and environmental sustainability (e.g. Cherp et al. 2011; Florini and Sovacool 2011; Karlsson-Vinkhuyzen et al. 2012). Finally, the renewable energy subfield is still dominated by national policy making as nation states continue to have sovereign control over the energy domain.

Various studies have introduced mappings of the institutional complexes for climate change (e.g. Keohane and Victor 2011; Abbott 2012; Widerberg et al.

2016) and energy (e.g. Sovacool and Florini 2012; Wilson 2015; Sanderink et al. 2018), but only a few zoom in on the institutional complex for renewable energy specifically. This is regrettable, as the subfield is most prominent within the climate-energy nexus and can be characterized as institutionally diverse (see Chapter 3). Furthermore, existing mappings are biased toward public institutions, mostly excluding nongovernmental institutions (e.g. Suding and Lempp 2007; Barnsley and Ahn 2014; Röhrkasten 2015). However, recent literature has argued that the global energy transition is driven by bottom-up and polycentric governance rather than through integrated international cooperation (Aklin and Urpelainen 2018; Meckling 2018). Hence, novel insights are needed into the institutional constellations and dynamics within the renewable energy subfield, to ultimately answer the guiding question: is the institutional complex of renewable energy contributing to the global energy transition in an effective manner?

A first step in the search for an answer to this question is to advance our understanding of the institutional complex and to evaluate coherence and management (as laid out in further detail in Chapter 2). First, coherence is understood as the harmony of institutional features and interactions across institutions toward an overarching purpose. Meso-level coherence, i.e. the level of coherence in the subfield as a whole, is determined based on the following indicators: first, the convergence/divergence among interpretations of the core norm, i.e. to substantially increase the share of renewables; second, the distribution of membership, i.e. a limited or wide range of targeted actors; and third, an (un)balanced allocation of governance functions. Micro-level coherence, i.e. the level of coherence between specific individual institutions, is assessed along the same three dimensions and, more importantly, mechanisms of interactions. These can be distinguished as cognitive (i.e. when knowledge is transferred), normative (i.e. when rules and norms interact), and behavioural (i.e. when impacts of behaviour intersect). Second, management is defined as attempts to deliberately steer interactions between two or more institutions (Zelli 2010), and is merely assessed at the micro level. It is determined based on the levels and agents (e.g. unilaterally or jointly), and consequences of management (e.g. increased harmony of institutional features: core norm, membership, governance functions, interaction mechanisms). Juxtaposing levels of coherence and management enables the characterization of the renewable energy subfield by synergy, coexistence/duplication, or conflict, or rather by division of labour, coordination, or competition (see Chapter 2).

For studying coherence and management at the micro level, three important multi-stakeholder partnerships were selected, since global (renewable) energy governance remains underdeveloped under the umbrella of the United Nations (UN) and in the intergovernmental sphere in general (Karlsson-Vinkhuyzen 2010; Röhrkasten 2015). Moreover, the chapter aims to go beyond dyadic relationships between intergovernmental institutions and seeks to analyze the plethora of interconnections among different forms of governance. The selected institutions include the Renewable Energy and Energy Efficiency Partnership (REEEP), the Renewable Energy Policy Network for the 21st Century (REN21), and Sustainable Energy for All (SEforALL), all of which can be described as transboundary cooperation efforts between public and private actors that aim at addressing a public policy objective (Schäferhoff et al. 2009; Pattberg and Widerberg 2014).

The analysis builds on three methodological steps: first, a thorough analysis of the institutional constellations, second, a content analysis of official documents and reports of the selected institutions, and third, an analysis of the views of climate and energy experts obtained through semi-structured interviews. The interviewees were staff members of the selected institutions, experts from academia and civil society organizations (CSOs), and government officials, who are closely associated with the respective institutions. Based thereon, this chapter advances our understanding of institutional complexity, specifically for global renewable energy governance. Therewith, it provides insights from which lessons can be drawn for governing the overall climate-energy nexus.

The chapter is structured as follows. Section 4.2 briefly introduces the topic of renewable energy and its centrality. Subsequently, Section 4.3 analyzes the institutional features and measures meso-level coherence for the renewable energy subfield as a whole. Thereafter, Section 4.4 determines micro-level coherence by examining institutional features and interaction mechanisms across the selected partnerships. Finally, Section 4.5 describes attempts to manage these interactions at the micro level, after which Section 4.6 concludes with some final remarks and suggestions for future research.

4.2 Renewable Energy: Providing Sustainable Energy for All

'*We all know that renewable energy is limitless and will last forever*' is what former UN Secretary-General Ban Ki-Moon stated in 2016 at the International Renewable Energy Agency (IRENA) Debate in Abu Dhabi.[1] This statement mirrors the high importance of the role of renewable energy in the world's trajectory to sustainable development.

Firstly, renewable energy is crucial for satisfying the increasing energy demand. In light of the world's growing population, energy security is a high priority for governments worldwide (Dubash and Florini 2011; Van de Graaf 2013), in the way that they wish to ensure an '*uninterrupted availability of energy sources at an*

[1] For the full statement: www.un.org/sg/en/content/sg/statement/2016-01-17/secretary-generals-remarks-international-renewable-energy-agency.

affordable price'.[2] At present this is particularly challenging, since finite energy sources are depleted rapidly, while global energy demand is rising sharply. As a consequence, diversification of energy sources is of great necessity, and renewables can play an important role in this. Solar, wind, and other types of renewable energy have the potential to alleviate the increasing scarcity, to decentralize the production of energy, and to diversify energy supply (Röhrkasten 2015).

Secondly, renewables are key to ensuring worldwide energy access. The challenge of energy access is related to the 1.1 billion people who do not have access to electricity and to the 2.8 billion people who continue to rely on biomass, coal, and kerosene for cooking (OECD/IEA 2017). Not only does this deprive this large part of the human population from economic modernization, it also poses urgent health threats and environmental degradation risks (Dubash and Florini 2011). This demonstrates the urgency to tackle the widespread and persistent lack of access to modern energy services, which is predominantly the case in rural areas in the developing world. Switching to renewables does not only reduce the indoor air pollution and improve the population's health, it is also highly suitable for small-scale and decentralized deployment, which is particularly important to address energy access (Röhrkasten 2015).

Thirdly, an increased uptake contributes to tackling the negative environmental externalities that are associated with today's energy systems, and the first and foremost issue related to energy is climate change. Other urgent environmental issues are air pollution, acid rain, contamination of marine environment, nuclear meltdowns, collapsed coal mines, natural gas explosions, dam breaches, and so forth (Dubash and Florini 2011; Florini and Sovacool 2011; and Röhrkasten 2015). This makes the imperative to substitute fossil fuels and further diversify the energy mix even stronger, and renewables have the potential to do so. However, an increased uptake has its own environmental risks. For example, the cultivation of biofuel crops is associated with soil degradation and deforestation; the construction of hydropower dams with disruption of local fish stocks; the use of nuclear energy with the danger of toxic substances; and the production of solar and wind energy with the displacement of food production, interventions in stability of ecosystems, and dangers to bird life (Van de Graaf 2013; Röhrkasten 2015).

4.3 Meso-Level Coherence

Before zooming in on institutions at the micro level, this section describes meso-level coherence for the overall subfield of renewable energy by describing its emergence, the core norm, membership, and governance functions.

[2] Definition of energy security derived from: www.iea.org/topics/energysecurity/.

4.3.1 Emergence of the Institutional Complex on Renewable Energy

The timeline in Figure 4.1 illustrates the emergence of the institutional complex for renewable energy. As of January 2017, the institutional complex consists of forty-six institutions with different constitutive characteristics. Even though first global environmental concerns were already raised in the 1970s, and the dependence on fossil fuels was already questioned in the Brandt Report (1980), it took until the early 1990s for interest in renewable energy to grow significantly.

Institutions that were established prior thereto mostly include intergovernmental cooperation efforts that were initially shaped by energy security concerns as a consequence of the oil crises in the 1970s. For example, the International Energy Agency (IEA), which initially focused on fossil sources of energy, slowly but surely widened its portfolio and extended its analyses to renewable energy (Van de Graaf 2012; Heubaum and Biermann 2015). The emergence of institutions from the early 1990s appears to be linked to the UN Framework Convention on Climate Change (UNFCCC), adopted in 1992, followed by the Kyoto Protocol in 1997. As an illustration, the same year the UNFCCC was adopted the Global Sustainability Electricity Partnership (GSEP) was established to decarbonize the world through sustainable electrification (GSEP 2018), and similar to the Kyoto Protocol, the CarbonNeutral Protocol (CNP) was set up in 1997 to stimulate carbon reductions, for example through renewable energy certificates (CNP 2018).

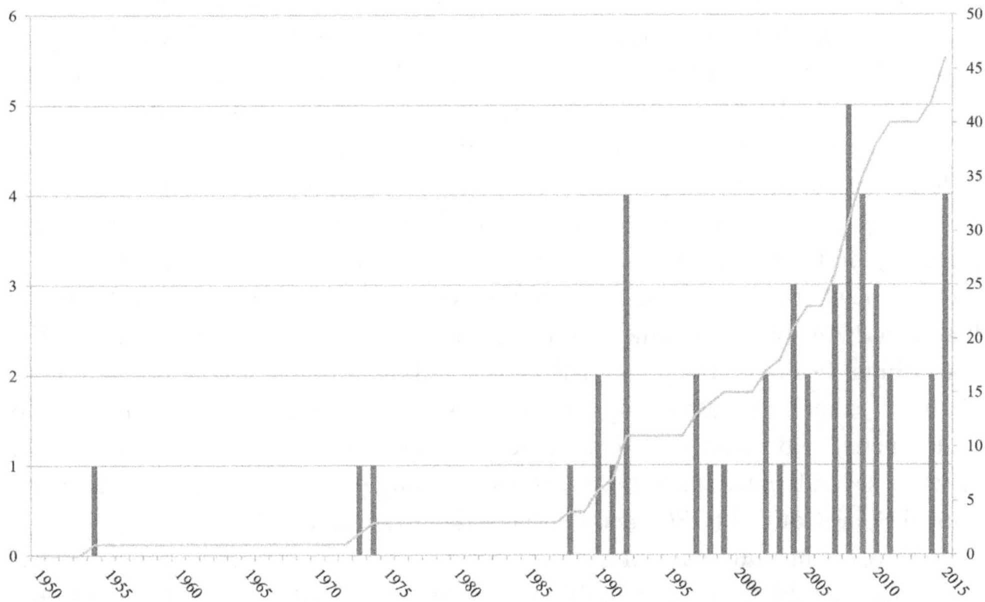

Figure 4.1 Starting years of renewable energy institutions from 1950 to 2015 (Author's data).

However, it took until the turn of the millennium for institutions to exclusively focus on renewable energy. In 2001 the topic was for the first time discussed at the UN's high political level, at the Ninth Session of the Commission on Sustainable Development (CSD) (Karlsson-Vinkhuyzen et al. 2012), although no substantial agreement was reached (Röhrkasten 2015). Instead, intergovernmental institutions started to emerge outside the UN system. For example, IRENA was established in 2009. It serves as a principal forum for transboundary cooperation and provides a repository of policy, technology, resource, and financial knowledge (IRENA 2018b). That same year, the Clean Energy Ministerial (CEM) was initiated, bringing together ministers with responsibility for clean energy, to promote policies and programmes, and share knowledge and best practices (CEM 2018).

A decade later, in 2015, a growing consensus at the UN level on the strong link between energy and poverty eradication eventually led to the inclusion of Sustainable Development Goal (SDG) 7 in Agenda 2030 to *'ensure access to affordable, reliable, sustainable and modern energy for all'* (United Nations 2015, 19). More importantly, SDG 7 included target 7.2 that commits countries to, *'by 2030, increase substantially the share of renewable energy in the global energy mix'* (United Nations 2015, 19). Along with the Paris Agreement and its target to keep global temperature rise well below 2 degrees Celsius, SDG 7 at least marks the emergence of universal objective for global (renewable) energy governance.

In parallel to this development, a somewhat smaller expansion of institutions took place. In 2011, for instance, the Low-Emissions Development Strategies Global Partnership (LEDS_GP) was established to facilitate learning, technical cooperation, and information exchange supporting low emission development strategies (LEDS_GP 2018). Furthermore, the Africa Renewable Energy Initiative (AREI) was initiated in 2015 to accelerate and harness the African continent's renewable energy potential (AREI 2018). While these two institutions focus primarily on the deployment of renewables to expand energy access, other institutions focus specifically on emissions reductions. For instance, RE100, established in 2014, brings together influential businesses to collectively promote the compelling business case for renewables (RE100 2018). Likewise, the Low Carbon Technology Partnerships initiative (LCTPI) was set up that same year to unite energy and technology companies to scale up renewables (LCTPi 2018). Finally, several other institutions exclusively target solar energy, including the Global Solar Council (GSC) and Global Solar Alliance (GSA).

In sum, the institutional complex for renewable energy comprises a multitude of institutions established within different contexts and with different institutional characteristics. The following subsections will further elaborate on some of these institutional characteristics and the variation across them.

4.3.2 The Core Norm of Renewable Energy

The Paris Agreement and the inclusion of SDG 7 as part of Agenda 2030 arguably constitute the major institutional incentive to ensure access to sustainable energy for all. More specifically, target 7.2 of SDG 7 sets the objective to substantially increase the share of renewable energy in the global energy mix by 2030 (United Nations 2015). Altogether, these institutional targets speak to the three critical challenges for global (renewable) energy governance: energy security, energy access, and environmental sustainability. In sum, the core norm for the renewable energy subfield can be described as: *to substantially increase the share of renewable energy in the global energy mix, in order to ensure access to and availability of clean energy for all.* The normative coherence of the renewable energy subfield depends on the degree to which this core norm is shared or disputed across institutions.

A closer inspection of the institutions' web pages and mission statements shows that, for one-third of the institutions (17 out of 46), this core norm applies literally. These are mostly public institutions, for instance IRENA and CEM, but also three private institutions, including GSEP, EUROSOLAR, and the International Solar Energy Society (ISES), as well as two multi-stakeholder partnerships, the Global Bioenergy Partnership (GBEP) and REN21. This implies that the majority of institutions interpret the core norm more selectively, by prioritising either one or two of the main objectives, i.e. energy security, energy access, and environmental sustainability, rather than approaching them in an integrated manner.

Sixteen institutions put environmental sustainability first, and interpret the core norm as to substantially increase the share of renewables to mitigate environmental externalities, most urgently the effects on the global climate. More specifically, seven of these institutions explicitly refer to the norms and targets set by the UNFCCC and its Kyoto Protocol. For instance, the LCTPI Rescale programme aims to scale up renewables and calls for action to *'stay below 2°C of global warming'* (LCTPi 2018). Most of the institutions with a focus on environmental sustainability are public, such as Regions 20 (R20), which supports subnational governments to develop low-carbon infrastructure projects, and the Climate Investment Fund (CIF), which provides investment programmes to scale up renewable energy in low-income countries (R20 2018; CIF 2018). In addition, there are several private institutions, such as RE100 and CNP, and multi-stakeholder partnerships, namely the Roundtable on Sustainable Biomaterials (RSB), LEDS_GP, and LCTPI.

Five institutions adhere to the core norm to substantially increase the share of renewables to expand access to clean energy sources, for the purpose of improving energy access as well as mitigating climate change. These include two public institutions, namely UN Energy (UN_EN) and the Clean Energy Solutions Centre (CESC), one private institution, the Gold Standard (GS), and two multi-stakeholder partnerships, namely REEEP and SEforALL. Finally, the remaining eight institutions target either energy security, energy access, or both, or energy security and environmental sustainability simultaneously.

Figure 4.2 provides an overview of the institutions according to the core norm they predominantly adhere to. Altogether, and allowing overlaps, forty-one institutions prioritize climate change mitigation, twenty-seven addressing energy access and twenty-two energy security policy objectives. In other words, the uptake of renewables is predominantly linked to mitigating climate change, while particularly the potential of renewables to address energy security concerns is less institutionalized.

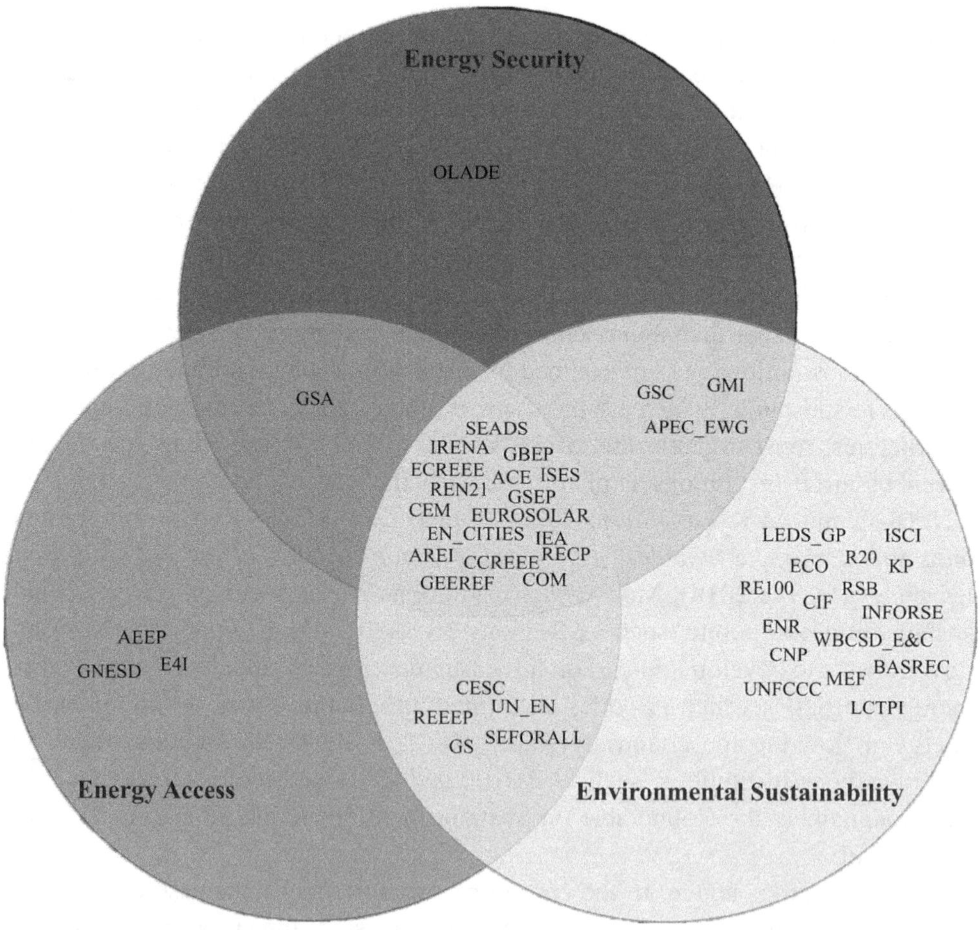

Figure 4.2 Distribution of renewable energy institutions according to their interpretation of the core norm (Author's data).

As a consequence of these diversified priorities, there is much room for trade-offs and potential conflicts. Institutions prioritising energy security and access may, in addition to renewables, turn to the more affordable energy sources that might have a negative impact on the environment (Röhrkasten 2015).[3] Additionally, expanding energy access implies increased energy demand, which in turn puts those institutions under pressure that seek to ensure energy security and environmental sustainability (Newell et al. 2011). On top of this, the subfield lacks a clear definition of what constitutes a renewable source of energy, resulting in frequent controversies on bioenergy, hydropower, and nuclear energy (e.g. Elliott 2000; Frey et al. 2002; Adamantiades and Kessides 2009).[4] While solar and wind are widely acknowledged as renewable energy sources within the renewable energy subfield, twenty-five institutions include bioenergy, seventeen (small-scale) hydropower, and three nuclear power.

These diverging views have entailed conflicts of interests and competition over resources, visibility, and media attention across institutions targeting these different energy sources.[5] As long as the potential of renewable energy to address energy security is not fully institutionalized, and as long as there is no consensus on what constitutes a renewable energy source, full substitution of fossil fuels by renewables, particularly in industrialized countries, may remain unattainable.

4.3.3 Membership

The governance triangle introduced in Chapter 3 distinguishes different types of institutions and actors that are involved in promoting the uptake of renewables globally. Figure 4.3 additionally summarizes the respective figures in a table.

The largest share of renewable energy institutions (28) are public. These include international organizations, such as IRENA and the IEA, as well as regional alliances, such as the Latin American Energy Organization (OLADE), the Asia-Pacific Economic Cooperation Energy Working Group (APEC_EWG), the Association of South East Asian Nations (ASEAN) Centre for Energy, and the Baltic Sea Region Energy Cooperation (BASREC). Besides these intergovernmental efforts, a number of institutions unite cities and regions in their search for

[3] Interview with Professor Thijs Van de Graaf, Ghent Institute for International Studies, Ghent University, 13 July 2018.

[4] Interview with Professor Thijs Van de Graaf, Ghent Institute for International Studies, Ghent University, 13 July 2018.

[5] Interview with Stefan Gsänger, Secretary-General, the World Wind Energy Association, and Vice Chair, Renewable Energy Policy Network for the 21st Century (REN21), 9 May 2018; interview with Benjamin Sovacool, Professor of Energy Policy, Science Policy Research Unit, University of Sussex, 10 May 2018; and interview with Stephan Singer, Senior Advisor Global Energy Policies at Climate Action Network International, 10 May 2018.

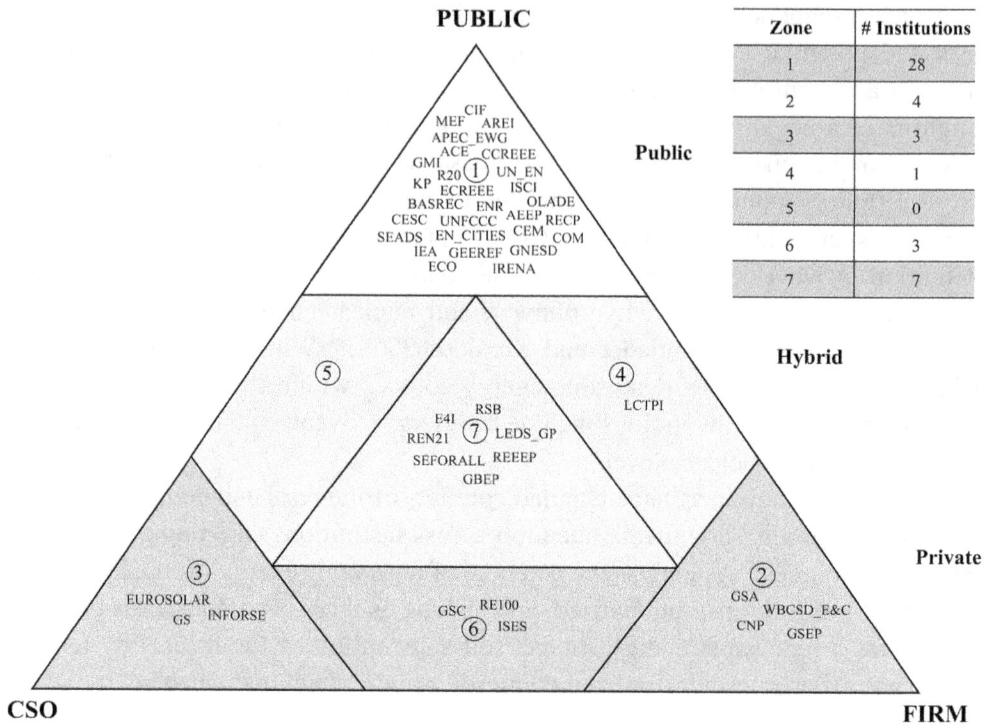

Zone	# Institutions
1	28
2	4
3	3
4	1
5	0
6	3
7	7

Figure 4.3 Distribution institutions per zone in the governance triangle (Author's data; see also Chapter 3).

appropriate strategies toward an energy transition, such as the International Solar Cities Initiative (ISCI), Energy Cities (EN_CITIES), the Covenant of Mayors (COM), and R20. While the influence of these different institutions is limited to their respective members, several other public institutions aim at assisting developing countries in increasing their share of renewables. Such institutions include, for example, the Africa-European Union Energy Partnership (AEEP) and the Global Energy Efficiency and Renewable Energy Fund (GEEREF).

In addition to public institutions, there are ten private institutions, of which four exclusively bring together firms, industry associations, and investors. One of these is the Energy and Climate Cluster of the World Business Council on Sustainable Development (WBCSD_E&C), which unites companies from different sectors to scale up climate and renewable energy solutions globally (WBCSD 2018). In addition, there are three institutions that exclusively include nongovernmental organizations (NGOs) and other organizations representing civil society. These include the Gold Standard (GS), set up by the World Wide Fund for Nature (WWF) and two other NGOs, which provides a standard for climate and development projects under the UNFCCC Clean Development Mechanism (Gold Standard

2018). Furthermore, three institutions include firms, industry associations, and investors, as well as NGOs and other civil society actors, e.g. GSC and ISES.

Finally, there are eight hybrid institutions, or multi-stakeholder partnerships, which represent collaborations across societal sectors. The LCPTI is the only institution restricting its membership to public actors and firms, industry associations, and investors, while the other seven bring together all three main types of actors. Three out of these seven partnerships have been selected for further analysis in Section 4.4 on micro-level coherence: REEEP, which develops financing mechanisms to strengthen markets for clean energy in low- and middle-income countries; REN21, which connects stakeholders to facilitate knowledge exchange and policy development toward a transition to renewable energy; and SEforALL, which marshals evidence, benchmarks progress, and connects stakeholders toward achieving SDG 7 and the Paris Agreement.

While the majority of renewable energy institutions are public, private and multi-stakeholder institutions have started to play a significant role in promoting a worldwide uptake of renewable energy. In other words, there is a wide variety of actors involved, ranging from governments, international organizations, cities, and subnational authorities, to companies, financial institutions, and not-for-profit organizations. Mapping and understanding these variations is an important step toward assessing not only the coherence but also the effectiveness of the institutional complex of this subfield, since collaborations between these actors are considered key for a successful global governance of renewables (e.g. Sovacool 2013).

4.3.4 Governance Functions

Renewable energy institutions perform different governance functions, which can be distinguished as setting 'standards and commitments', 'operational' activities, sharing 'information and networking', and 'financing' (see Chapter 2). Ideally, all of these governance functions are performed in a complementary manner, in accordance to the core norm, and without functional gaps or duplications.

The largest share of institutions in the sample (17 out of 46) governs renewable energy through 'information and networking', in combination with 'operational' activities. This implies that most institutions combine evaluation activities, as well as collecting and publishing information, with pilot projects, technical assistance, and capacity building. For example, Energy for Impact (E4I) assists local businesses and project developers in East and West Africa to expand energy access and publishes reports on related topics (E4I 2018). In addition, fifteen institutions exclusively govern through 'information and networking'. A well-known example is IRENA as an international organization, which claims to serve as a centre of excellence, and a repository of knowledge (IRENA 2018b). This said, several multi-stakeholder partnerships also fulfil such a role, for instance REN21, which connects key stakeholders to facilitate knowledge exchange (REN21 2018c).

Furthermore, there are eight institutions that set 'standards and commitments', and more specifically develop rule-making processes, mandatory or voluntary commitments, or schemes for implementation and enforcement. This governance function is not merely reserved for public institutions, since various private and multi-stakeholder institutions provide certifications and standards to which different actors can voluntarily commit. For instance, whereas the Kyoto Protocol sets binding emission-reduction targets (United Nations 1998), the Roundtable on Sustainable Biomaterials (RSB) sets principles and criteria to help operators, brand owners, and investors to identify and manage sustainability issues (RSB 2018). Finally, five institutions are involved in 'financing' to promote a global uptake of renewables, for instance through funding projects or developing aid programmes. These solely include public institutions, with one exception: REEEP, which '*invests in clean energy markets in developing countries to reduce CO_2 emissions and build prosperity*' (REEEP 2018b).

Summarizing this and the previous subsection, Table 4.1 provides an overview of the individual renewable energy institutions based on their governance functions and membership. Altogether, there is not a clear division of labour in terms of governance functions across the public, private, and hybrid institutions in the renewable energy subfield. Yet, while private and hybrid institutions play a role in all governance functions, the distribution of institutions in Table 4.1 clearly conveys the dominance of public institutions. Moreover, the table shows that there is a certain profusion of information-sharing and networking opportunities, while standards and commitments, and financing mechanisms, are limited to a few institutions. This suggests that, within the renewable energy subfield, soft measures prevail over hard ones.

4.3.5 Summary: Coherence at the Meso Level

The renewable energy subfield is institutionally complex in various ways. It includes a high number of institutions with different constitutive characteristics, covers several sources of energy and distinctive technologies, and targets no less than three critical challenges. The question remains, however, whether this plurality of institutions and objectives affects the normative, functional, and membership-related coherence in this subfield.

Altogether, this meso-level coherence, i.e. the level of coherence in the subfield as a whole, can be determined as low to medium, based on three conclusions. First, while there exists a core norm to substantially increase the share of renewables, the

Table 4.1 *Overview of governance functions across different types of institutions (public, hybrid, private) for renewable energy (Author's data).*

	Public	Hybrid	Private
Standards & Commitments	CoM, KP, UNFCCC	RSB, SEforALL	CNP, GS, RE100
Operational	BASREC		
Information & Networking	IRENA, ISCI, MEF, ACE, APEC_EWG, ECO, EnR, SEADS, CESC	LCTPi, REN21, LEDS_GP	GSA, INFORSE, ISES
Financing	CIF, GEEREF	REEEP	
Standards & Commitments; Operational			
Operational; Information & Networking	CEM, GMI, R20, AEEP, CCREEE, ECREEE, EN_CITIES, GNESD, IEA, OLADE, UN_EN	GBEP, E4I	GSEP, WBCSD_EC, EUROSOLAR, GSC
Information & Networking; Financing	RECP		
Standards & Commitments; Information & Networking			
Standards & Commitments; Financing			
Operational; Financing	AREI		

majority of institutions have diverging views on which objectives to prioritize. Most of these focus on promoting renewables for the purpose of mitigating climate change, or, to a lesser extent, to expand energy access, while a minority of institutions targets the potential of renewable energy to ensure energy security. As a consequence of this divergence, there is much room for trade-offs, controversies, and potential conflicts among institutions and actors across them, suggesting that normative coherence is low.

Second, the renewable energy subfield is currently dominated by public institutions, including international organizations as well as regional and subnational institutions. At the same time, the number of private institutions and multi-stakeholder partnerships has been steadily growing. The range of actors involved in the sample of this study thus increasingly stretches from public to private, implying medium membership-based coherence.

Third, all governance functions are covered by one or more institutions: most govern renewables through information-sharing and networking, and a fair share of institutions are involved in operational activities. By contrast, standards and commitments are mostly set in the form of voluntary standards and certification schemes, and financing schemes are provided by only a few institutions. In other words, there appears to be a profusion of informal activities at the expense of other important governance functions, which suggests medium functional coherence.

4.4 Micro-Level Coherence

As explained in Chapter 2, the assessments of the subfields do not stop at looking at the core norms, membership, and governance functions across the subfields as a whole. Additionally, the case studies in this volume scrutinize relations between individual institutions, i.e. at the micro-level. Whereas previous studies mostly focus on dyadic relationships between intergovernmental institutions (e.g. Charnovitz 2003; Oberthür and Gehring 2006b; Zelli and van Asselt 2010), this study analyzes a plethora of interconnections among different types of institutions. This allows for a much more encompassing assessment of an entire subfield, especially one so densely populated as renewable energy.

The subfield comprises no less than forty-six institutions, and three specific ones were selected for the in-depth analysis presented in this chapter. The following subsections describe the institutions under scrutiny, and determine micro-level coherence based on core norm, membership, and governance functions, and more importantly, by identifying mechanisms of interaction among the selected institutions.

4.4.1 Institutions under Scrutiny

Even though the majority of renewable energy institutions are public, private and multi-stakeholder institutions play an important role in governing renewables. This is particularly true for multi-stakeholder partnerships, which bring together public and private actors to collectively contribute to a public policy goal. Such partnerships generally emerge as a response to a lack of intergovernmental cooperation (Szulecki et al. 2011, 713). As national policy-making continues to dominate when it comes to energy issues, international energy governance is weakly developed, and especially so for renewable energy. Furthermore, cooperation between public and private actors is considered of great importance for increasing the share of renewables in the global energy mix (Sovacool 2013). Whereas previous literature focused predominantly on international organizations for energy (e.g. Colgan et al. 2012; Leal-Arcas and Filis 2013; Wilson 2015), a shift toward multi-stakeholder partnerships can provide novel insights on the interactional dynamics within the overall institutional complex for renewable energy.

The Renewable Energy and Energy Efficiency Partnership (REEEP) was one of the first multi-stakeholder partnerships on energy-related issues. Led by the British government, a group of regulators, businesses, and NGOs announced REEEP in 2002 at the World Summit on Sustainable Development in Johannesburg (Florini and Sovacool 2009; Röhrkasten 2015). Two years later the partnership was formally established as an international NGO based in Vienna. By investing in clean markets and targeting small- and medium-sized enterprises, REEEP aims to accelerate market-based deployment of renewable energy and energy-efficient systems in low- and middle-income countries (REEEP 2018b). The partnership relies on donors, which include governments, multilateral and international organizations, NGOs, and foundations. More than 350 members currently back up REEEP, including businesses, NGOs, national governments, research institutes, and many other entities. The partnership is governed by a Governing Board and an Advisory Board and is steered by an international team with more than twenty staff members and consultants. REEEP is seen as an important multi-stakeholder partnership with a clear purpose, significant output, and strong institutional formality (Pattberg et al. 2009; Szulecki et al. 2011; Sovacool and Van de Graaf 2018).

Besides REEEP, the Renewable Energy Policy Network for the 21st Century (REN21) forms an important coalition of different stakeholders to advance renewable energy policy. The German government initiated REN21 at the International Conference for Renewable Energies in 2004 in Bonn, after which it was formally launched in Copenhagen in June 2005 (REN21 2005; Röhrkasten 2015). The partnership brings together different stakeholders to facilitate knowledge exchange, policy development, and joint action toward a rapid global transition to renewable energy (REN21 2018c). Its existence depends on grants offered by governments, international organizations, and other donors, and by the end of 2017, REN21 counted sixty-four members, including industry associations, international organizations, NGOs, national governments, and research entities. The partnership is governed by its Bureau, General Assembly, and Steering Committee and has a small secretariat housed at the UN Environment Programme (UNEP) in Paris. It is considered as an important advocacy network and global governor for renewable energy (Szulecki et al. 2011; Röhrkasten 2015).

A more recently established multi-stakeholder partnership is Sustainable Energy for All (SEforALL). It was initially launched as a UN initiative by former UN

Secretary-General Ban Ki-Moon in 2011, and thereafter formalized as a non-profit quasi-international organization (Röhrkasten 2015). While SEforALL inherited close ties to various UN agencies, it is now open to different stakeholders including governments, businesses, financiers, development banks, communities, and others. SEforALL's mission is threefold: to ensure universal access to modern energy services; to double the global rate of improvement in energy efficiency; and to double the share of renewable energy in the global energy mix (SEforALL 2018a). The partnership relies on donor contributions mostly coming from national governments, and in 2017 SEforALL counted more than eighty partners. These can be distinguished among funding partners, delivery partners (partners that commit to contribute quantifiable results and to report on progress), proud partners (partners that support SEforALL's objectives, and use the name and platform to amplify meaningful work), and those participating in SEforALL's regional and thematic hubs, and accelerators (SEforALL 2018b). The partnership is governed by an Administrative Board and a Funder's Council, which guide SEforALL's Global Team that is headquartered in Vienna and partly operating in Washington, DC. With this, SEforALL has a unique standing in the institutional complex, since it is an important multi-stakeholder partnership and major UN initiative at the same time (Röhrkasten and Westphal 2013).

4.4.2 Interlinkages

The three selected multi-stakeholder partnerships show differences as well as commonalities with regard to their institutional features.

First, the interpretations of the core norm for renewable energy among the three multi-stakeholder partnerships largely overlap. Particularly the partnerships REEEP and SEforALL commonly adhere to a core norm strongly influenced by targets set by the UNFCCC regime and SDG 7: to substantially increase the share of renewables for universal energy access and to limit global warming to 2 degrees Celsius. For example, REEEP repeatedly stresses the importance to connect its goals, targets, and metrics to the Paris Agreement (REEEP 2016a), and celebrates the inclusion of SDG 7 as a validation of REEEP's work over the years to expand energy access (REEEP 2016a, 11). Similarly, SEforALL's objectives include to ensure universal access to energy, double the global rate of energy efficiency, and double the share of renewables, which were formulated in the run-up to SDG 7. In addition, SEforALL reiterates that actions to achieve these objectives should be in line with the 2 degrees target agreed upon in the Paris Agreement (SEforALL 2018c). By contrast, REN21 takes on a broader approach to the core norm. REN21's flagship publication, the Renewable Energy Global Status Report (GSR) 2018, acknowledges that scaling up renewables is crucial for limiting temperature rise below 2 degrees and for meeting the aspirations of SDG 7 (REN21 2018b). However, the partnership additionally takes into consideration the policy objective to boost national energy security (REN21 2018b).

Second, the membership directories of the three partnerships partly overlap. For instance, REEEP and REN21 share sixteen members. These include international organizations, such as the European Commission and the UNEP, and national governments such as Brazil, Germany, and the United Kingdom. In addition, REEEP and REN21 share various NGOs as members, e.g. the WWF and The Energy and Resources Institute (TERI). As SEforALL is essentially a partnership between the UN and the World Bank, there are no shared members with REEEP and REN21. However, some of the funding and delivery partners of SEforALL are members to REEEP (9), including respectively Austria, the European Commission, and Germany, as well as Johnson Control, UNEP, and the UN Foundation (SEforALL 2018b). Similarly, a set of funding and delivery partners are members to REN21 (6), including respectively Germany, the European Commission, Norway, and the United Kingdom, as well as the Global Association for Off-grid Solar Energy Industry (GOGLA) and UNEP (SEforALL 2018b). Among approximately 500 members spread across the three partnerships, there are only five actors all three have in common: the European Commission, UNEP, Germany, Norway, and the United Kingdom.

Third, when it comes to governance functions, the selected partnerships do not overlap, but rather complement each other. However, it is important to note that the three partnerships were also selected according to their variations in governance functions. Whereas REEEP develops and provides financing mechanisms to advance market readiness for clean energy services in low- and middle-countries (financing) (REEEP 2018b), REN21 connects different stakeholders to facilitate knowledge exchange toward a rapid transition to renewable energy (information and networking) (REN21 2018c). Finally, SEforALL connects leadership to mobilize action on SDG 7 specifically (standards and commitments) (SEforALL 2018c). Consequently, the three partnerships also target different entities; while SEforALL speaks to government leaders and REN21 to policy makers more broadly, REEEP targets small- and medium-sized enterprises.

4.4.3 Mechanisms

Institutional interactions can be broadly understood as situations in which the policy processes, knowledge, norms, or functions of two or more institutions are connected, and affect the development, performance, and impact of these institutions (Oberthür and Gehring 2006a; Zelli et al. 2012). Hence, in addition to drawing parallels based on core norm, membership, and governance function, it is key to examine the underlying interaction mechanisms between and beyond the selected partnerships. The following subsections distinguish and describe these as cognitive, normative, and behavioural (see Chapter 2; Stokke 2001; Oberthür and Gehring 2006a).

4.4.3.1 Cognitive

A cognitive interaction is driven by the power of knowledge and persuasion, and can be seen as cross-institutional learning (Stokke 2001; Oberthür and Gehring 2006a). In other words, a cognitive interaction can be determined when knowledge and information are exchanged, or certain practices and methodologies are transferred from one to the other institution. As there is a large share of institutions that govern renewable energy through information and networking, there is presumably a high degree of cognitive interaction across the institutional complex.

For REEEP, various instances of cognitive interactions were found. First, REEEP applies a framework developed by the World Bank in 2015 for SEforALL to define the concept and measures of energy access (World Bank 2015; REEEP 2018a). Second, REEEP's regionally focused report on Powering India is informed by the IEA's India World Energy Outlook (WEO) 2015, and REEEP's report supporting a transition to inclusive green economies in African countries is influenced by IRENA's Africa 2030 report (REEEP 2016b, 2017). Vice versa, REEEP's publications have informed SEforALL and IRENA. For instance, the 'Making the Case' report published by REEEP influenced the Water-Energy-Food Nexus High Impact Opportunity (HIO) set up by SEforALL (REEEP 2015). The SEforALL HIOs serve as platforms that bring together stakeholders working on initiatives for the purpose of highly relevant topics related to clean energy, such as mini-grids and sustainable bioenergy. Additionally, REEEP's publication 'Making the Case' contributed to IRENA's report on 'Renewable Energy in the Water, Energy and Food Nexus' (REEEP 2015).

Cognitive interactions are found in larger numbers for REN21, since it is this institution's primary role to share information and set up networking opportunities. It is especially the work of IRENA and the IEA that is regularly cited in REN21's flagship GSRs (e.g. REN21 2017a; REN21 2018b). For instance, REN21's latest GSR features 85 pages of endnotes including no less than 386 references to the IEA's World Energy Outlooks, statistical reports, regional market analyses, and energy and CO_2 reports, and 161 references to IRENA's calculations and thematic reports (REN21 2018b). Similarly, REN21's regional status reports, such as those focused on the East African Community and the regions of the UN Economic Commission for Europe (UNECE), are influenced by SEforALL's data and information (REN21 2016a; REN21 2017b).

Besides, many more regionally focused and energy access-oriented institutions inform REN21's regional reports, for instance the Economic Community of West African States (ECOWAS) Centre for Renewable Energy and Energy Efficiency (ECREEE), the Africa-EU Renewable Energy Cooperation Program (RECP), CIF, and E4I. Vice versa, the information REN21 shares through its GSRs is widely acknowledged,[6] and regularly shared at key events. For example, a preview of GSR 2015 was presented at the IRENA Council that same year, GSR 2016 was launched at CEM 7 in San Francisco, and the Global Futures Report of 2017 was introduced at the 2017 SEforALL Forum (REN21 2018a).

Finally, SEforALL shows similar cognitive interactions, although to a lesser extent. Besides the SEforALL HIO being informed by REEEP, the statistics, data, and country profiles of the IEA and IRENA feed into SEforALL's Heat Maps. These inform the international community about which regions should be prioritized to close the energy access gap (SEforALL 2018b). Similarly, knowledge of ECREEE, AREI, CIF, WBCSD, and the Global Network on Energy for Sustainable Development (GNESD) has been included in SEforALL's publication on the state of electricity access worldwide (SEforALL 2017).

4.4.3.2 Normative

The normative type of interaction (or: interaction through commitment) occurs when the commitments, norms, and principles upheld by one institution confirm or contradict those of other institutions (Stokke 2001; Oberthür and Gehring 2006a). On the one hand, a low degree of normative interaction can be expected for the entire institutional complex, as the majority of institutions interpret the core norm more selectively by prioritizing either one or two of the core objectives, i.e. energy security, energy access, and environmental sustainability. On the other hand, Section 4.4.2 has shown that the selected partnerships are rather consentient in this regard, thus the normative interactions between REEEP, REN21, and SEforALL are considerable.

It is an enormous task to carefully compare the commitments, norms, and principles of REEEP, REN21, and SEforALL with those of all other institutions for renewable energy. However, some overlaps are expected based on the above broad evaluation of the institutions' core norm (see Section 4.3.2). First and foremost, all three partnerships show strong normative interaction with the

[6] Interview with Stefan Gsänger, Secretary-General, the World Wind Energy Association, and Vice Chair, Renewable Energy Policy Network for the 21st Century (REN21), 9 May 2018; interview with Stephan Singer, Senior Advisor Global Energy Policies at Climate Action Network International, 10 May 2018; and interview with Frank Van der Vleuten, Senior Energy Expert of the Climate Team at the Ministry of Foreign Affairs, the Netherlands, 5 June 2018.

UNFCCC regime. As described earlier, REEEP and REN21 as well as SEforALL stress the importance of aligning their commitments, norms, and principles with the 2 degrees target set by the Paris Agreement under the UNFCCC regime. In addition, the commitments, norms, and principles of REEEP and SEforALL necessarily overlap with those institutions similarly prioritizing renewable energy for environmental sustainability and energy access, including CESC, UN Energy, and the Gold Standard (GS). Likewise, REN21's commitments, norms, and principles presumably overlap with such institutions that are similarly inclusive toward energy security objectives, such as CEM, IRENA, and AREI.

4.4.3.3 Behavioural

A behavioural interaction refers to situations in which the actions undertaken by one institution, or members thereof, are supportive or disruptive for the performance of other institutions (Stokke 2001; Oberthür and Gehring 2006a). For instance, if an institution aims to expand access to energy services in rural areas by providing clean cooking appliances, these activities inherently support actions undertaken by institutions to foster emission reductions. In contrast, carbon offsetting programmes developed by an institution may undermine the efforts of institutions aiming at 100 per cent renewable energy. Thus, behavioural interactions can be driven by matching objectives (Gehring and Oberthür 2009), but also include, for instance, shaming, pressure, brand management, or monitoring each other's performances (see Chapter 2). For the scope of this study, it would lead too far to measure the actual impact of behaviours and activities, so the following analysis suffices with distinguishing and illustrating major behavioural interactions.

It is REEEP's main objective to advance clean energy services in low- and middle-income countries. Similarly, it is SEforALL's priority to secure affordable and reliable clean energy for all by 2030. Thus, synergistic behavioural interactions of REEEP and SEforALL most likely occur with institutions with matching objectives at the intersection of energy access and environmental sustainability. These include UN Energy, Gold Standard (GS), and CESC (see Section 4.3.2). The synergies are supposedly strong: first, since the membership directories of these institutions only partly overlap, so that the matching objectives apply to a wider range of actors, second, as these institutions pursue these objectives through different means of governance, i.e. governance functions, complementary to those of REEEP and SEforALL.

In addition, behavioural interactions through monitoring and potentially influencing the performance of other institutions take place between REEEP and SEforALL on the one hand, and the UNFCCC, IRENA, and CEM on the other. First, REEEP visits and actively participates in IRENA's General Assemblies, the yearly SEforALL Forum, and the UNFCCC COPs.[7] Second, SEforALL similarly performs sustainable energy diplomacy at the General Assemblies of IRENA, the UNFCCC COPs, and at the Clean Energy Ministerials (CEM) (SEforALL 2018b). Moreover, SEforALL is heavily involved in the UNFCCC process, particularly through Rachel Kyte, CEO of SEforALL and Special Representative of the UN secretary-general, and through organizing Energy Days jointly with IRENA at COP21 and 22, and presumably future COPs (SEforALL 2018d).

As mentioned earlier, it is REN21's mission to ensure a global transition to renewable energy, to limit temperature rise below 2 degrees, to meet the targets set by SDG 7, and to boost energy security. Hence, synergistic behavioural interactions of REN21 are expected with the sixteen remaining institutions with matching objectives at the intersection of energy security, energy access, and environmental sustainability (see Section 4.3.2). Similar to the interactions of REEEP and SEforALL with other institutions, these interactions of REN21 likely yield considerable benefits for both sides, since they expand the range of actors to which these objectives apply and cover complementary governance functions.

In addition, REN21 monitors and potentially influences the performance of the UNFCCC, IEA, IRENA, and SEforALL. The partnership actively participates in the UNFCCC COPs. For instance, in the run-up to COP21 in Paris, REN21 joined forces with the Covenant of Mayors (COM) to set up the Paris Process on Mobility and Climate (PPMC), and organized a series of events on 're-energising the future' together with IRENA (REN21 2015). Also at the following COPs in Marrakech and Bonn, REN21 hosted and participated in several renewable energy events. On top of that, REN21 regularly attends IRENA's General Assemblies, and SEforALL's yearly Forum, and is a member to the IEA's Renewables Industry Advisory Board and IRENA's Coalition for Action (IRENA 2018c; REN21 2015, 2016b).

4.4.4 Summary: Coherence at the Micro Level

While having in common that they are key governing institutions for renewable energy, the three selected multi-stakeholder partnerships are different in a variety of ways. Whereas REEEP is backed up by more than 350 members, REN21 and SEforALL 'only' have 64 and 86 members, respectively. In addition, while SEforALL speaks to government leaders and REN21 to policy makers more broadly, REEEP targets small- and medium-sized enterprises. Finally, REN21 provides policy-relevant information to support a global transition toward

[7] Interview with Katrin Harvey, Senior Manager, the Renewable Energy and Energy Efficiency Partnership (REEEP), 9 May 2018.

renewables, whereas SEforALL connects leadership to mobilize action on SDG 7, and REEEP mobilizes funding to accelerate market-based deployment of renewables.

In addition to these more obvious differences, a closer analysis of the institutional features and interaction mechanisms helped determine further aspects of micro-level coherence. First, the three institutions largely share their interpretations of the core norm, and the governance functions they perform are complementary, whereas the membership directories show little overlap. In other words, the normative, functional, and membership-based coherence and complementarity across the selected institutions is high.

Second, there is an abundance of cognitive interactions between the selected institutions and beyond, which substantiate the dominance of information and networking activities within the renewable energy subfield. In addition, there are considerable normative and behavioural interactions, resulting in various synergies while, at the same time, clustering different sets of institutions around certain priorities. The interaction mechanisms therefore appear to contribute to a functional imbalance and normative divergence in the subfield as a whole. In summary, despite the fact that the institutional features between the selected institutions are highly coherent, micro-level coherence as a whole should rather be qualified as medium.

4.5 Micro-Level Management

Finally, this section zooms in on deliberate attempts to manage institutional interactions among the renewable energy institutions analyzed in the previous section. These are deliberate attempts that seek to improve institutional interaction and its consequences, so as to prevent or strengthen the influence of one institution on the performance of another (Stokke 2001; Oberthür 2009). Typical examples are the provision of guiding principles by an overarching institution, joint coordination of activities across institutions, or unilateral management by individual institutions, for the purpose of more efficient goal-attainment (Oberthür 2009, 375–376). Such management attempts may lead to stronger coherence in terms of institutional features, for instance increased convergence among interpretations of the core norm, or novel interaction mechanisms for improved exchange processes.

First, the UNFCCC regime and Agenda 2030 come forward as important overarching frameworks for the three selected partnerships. As shown in Section 4.4.2, REEEP, REN21, and SEforALL ensure that their activities match the 2 degrees target of the Paris Agreement and SDG 7. For the subfield in general, these overarching goals provide 'international agreement', or at least a high degree of consensus, to globally phase out fossil fuels and foster a transition toward renewables.[8] This said, besides the three partnerships 'only' thirteen other institutions explicitly link their activities to the UNFCCC and SDG 7. This suggests that this overarching framework has not yet fully made its way to all renewable energy institutions.

Second, three examples were found of how interactions have been managed jointly by the three partnerships put under scrutiny here. First, REEEP and REN21 collectively operate reegle.info, which is a publicly recognized information portal on renewables, energy efficiency, and climate change.[9] The portal provides country energy profiles, energy statistics and research, and a directory of relevant stakeholders (REEEP 2018c). Second, REEEP worked with IRENA to create and launch the Renewables Tagger in 2016, which is a specialized version of the Climate Tagger and automatically scans and sorts data and documents holding renewable energy content to support knowledge-driven organizations to streamline their information (Climate Tagger 2018). Third, REN21, IEA, and IRENA have recently partnered up and published a report together on 'Renewable Energy Policies in a Time of Transition' (IRENA, OECD/IEA, and REN21 2018).

Finally, REN21 and SEforALL unilaterally manage institutional interactions with third institutions. First, REN21 and its flagship GSR, more specifically, facilitate numerous cognitive interactions.[10] All members to REN21 can contribute to the publication, and various institutions provide authors, contributors, and reviewers, such as IRENA and the IEA (REN21 2017a; REN21 2018b). Second, SEforALL provides an important overarching platform for various institutional interactions, particularly through its thematic and regional hubs and accelerators.[11] For instance, IRENA hosts the SEforALL thematic hub on renewable energy; REN21, CESC, and ECREEE take part in SEforALL's People-Centred Accelerator; and OLADE is an important player in SEforALL's regional hub for Latin America and the Caribbean. On top of that, SEforALL's Global Tracking

[8] Interview with Frank Van der Vleuten, Senior Energy Expert of the Climate Team at the Ministry of Foreign Affairs, the Netherlands, 5 June 2018; interview with Professor Thijs Van de Graaf, Ghent Institute for International Studies, Ghent University, 13 July 2018; and interview with Benedikt Hoskuldsson, Lead Partnership Specialist, Sustainable Energy for All (SEforALL), 21 September 2018.

[9] Interview with Stephan Singer, Senior Advisor Global Energy Policies at Climate Action Network International, 10 May 2018.

[10] Interview with Stefan Gsänger, Secretary-General, the World Wind Energy Association, and Vice Chair, Renewable Energy Policy Network for the 21st Century (REN21), 9 May 2018; interview with Stephan Singer, Senior Advisor, Global Energy Policies at Climate Action Network International, 10 May 2018; interview with Dr. Sybille Röhrkasten, Scientific Project Lead, Pathways to Sustainable Energy at the Institute for Advanced Sustainability Studies (IASS), 17 May 2018; and interview with Professor Thijs Van de Graaf, Ghent Institute for International Studies, Ghent University, 13 July 2018; and interview with Laura Williamson, Outreach and Communication Manager, Renewable Energy Policy Network for the 21st Century (REN21), 27 September 2018.

[11] Interview with Benedikt Hoskuldsson, Lead Partnership Specialist, Sustainable Energy for All (SEforALL), 21 September 2018.

Framework reports of 2015 and 2017, which measured progress on SDG 7, were coordinated by the IEA (World Bank and IEA/OECD 2015; World Bank and OECD/IEA 2017).

In sum, the renewable energy subfield is characterized by managed relationships, which, most significantly, provide an overarching normative framework and address cognitive interactions – and therewith, the potential overflow of information in the renewable energy subfield.

4.6 Conclusions

A global uptake of renewable energy is of paramount importance for a sustainable energy future, and while the share in the global energy mix is increasing, the growth rate is not sufficient to reach the targets set by SDG 7 in Agenda 2030 (United Nations 2018). Hence, effective global governance continues to play an important role in promoting renewables. As Chapter 3 has shown, global renewable energy governance is characterized by considerable institutional complexity. However, it is yet unclear whether this complexity significantly qualifies the institutional complex's impact on the global energy transition. To this end, this chapter scrutinized coherence and management within the renewable energy subfield, examining institutional features at the meso level, and interaction mechanisms and management attempts at the micro level.

The analysis of the subfield, comprising forty-six institutions, shed light on various important connections across institutions and their properties. First, while one-third of renewable energy institutions share the core norm to increase the proportion of renewables for energy security, energy access, and environmental sustainability, the majority of institutions interpret the core norm more selectively and prioritize either one or two of these objectives. Second, the subfield is dominated by public institutions, complemented by various private institutions and multi-stakeholder partnerships. Third, whereas most institutions facilitate information exchange and networking and, to some extent, implement projects on the ground, a significantly smaller set of institutions develops standards and commitments and financing mechanisms. Hence, the renewable energy subfield is characterized by diversified priorities – with a wide variety of institutions and actors, and with governance functions unevenly performed. The degree of meso-level coherence is therefore low to medium.

The selected multi-stakeholder partnerships, REEEP, REN21, and SEforALL, provided more detailed insights on interactional specifics at the micro-level. While these partnerships largely share the core norm for renewable energy, their membership directories hardly overlap, and governance functions are mostly complementary. Furthermore, cognitive interaction is the predominant mechanism involving the three partnerships put under scrutiny, notwithstanding the relevance of certain normative and behavioural interactions. Since these interaction mechanisms seem to aggravate the normative divergence and functional imbalance in the subfield, micro-level coherence can be determined as medium.

Besides interaction mechanisms, various management attempts were found to steer institutional interactions and foster synergies across renewable energy institutions. These mostly provide an overarching normative framework and manage the potential overflow of information within the subfield. Hence, this chapter concludes that, with a medium degree of coherence and management mechanisms in place, the renewable energy subfield is largely characterized by coordination (see Table 2.1, Chapter 2). However, such a densely populated subfield dealing with several critical energy challenges may require more than ad-hoc coordination to iron out controversies, trade-offs, and potential conflicts.

For the subfield to move toward a stronger division of labour, i.e. deliberate and continuous sharing of governance functions and norms for complementary membership (see Chapter 2), the following measures are recommended. First, the role of renewable energy to address energy security, energy access, and environmental sustainability in an integrated manner needs to be fully institutionalized. A reframing of the global energy challenge and the role of renewables may contribute to such a development (Sanderink 2019), as well as an expansion of membership of institutions toward those actors that are concerned with energy security. Second, the subfield would benefit from a track record or clearinghouse of the activities performed by existing renewable energy institutions, so that duplication and conflictive impacts can be resolved or prevented. Finally, it is necessary for institutions to strive for more cognitive alignment and some common understanding when it comes to defining renewable sources of energy. These three measures may require one coordinating institution, with IRENA being the likely choice: it advocates for a widespread adoption of renewables for energy security, energy access, and environmental sustainability; is closest to universal membership of all institutions in the subfield; and already positions itself as a principle platform for cooperation and repository of expertise (IRENA 2018b). Chapter 8 substantiates these policy recommendations in further detail.

Finally, this chapter gives rise to new questions that open important research opportunities. For example, how does the level of coherence and management relate to the effectiveness and legitimacy of individual institutions and the institutional complex for renewable energy as a whole? Or, what recommendations can be provided to specific public, civil society, or business actors that are trying to navigate the institutional complex? While this chapter provided a novel contribution on questions of coherence and management in renewable energy governance, these further questions will be revisited in Chapters 7 and 8.

4.7 References

Abbott, K. W. 2012. The Transnational Regime Complex for Climate Change. *Environment and Planning C: Government and Policy* 30(4), 571–590. https://doi.org/10.1068/c11127.

Ademantiades, A. and Kessides, I. 2009. Nuclear Power for Sustainable Development: Current Status and Future Prospects. *Energy Policy* 37(12), 5149–5166. https://doi.org/10.1016/j.enpol.2009.07.052.

Aklin, M. J. and Urpelainen, J. 2018. *Renewables: The Politics of a Global Energy Transition*. Cambridge, MA: MIT Press.

AREI (Africa Renewable Energy Initiative). 2018. About AREI. Available at www.arei.org/#about.

Barnsley, I. and Ahn, S. 2014. *Mapping Multilateral Collaboration on Low-Carbon Energy Technologies*. Paris: IEA.

CEM (Clean Energy Ministerial). 2018. About the Clean Energy Ministerial. Available at www.cleanenergyministerial.org/about-clean-energy-ministerial.

Charnovitz, S. 2003. Trade and Climate: Potential Conflicts and Synergies. In *Beyond Kyoto: Advancing the International Effort against Climate Change*, edited by J. A. Aldy, J. Ashton, R. Baron, D. Bodansky, S. Charnovitz, E. Diringer, T. C. Heller, et al. Arlington, VA: Pew Center on Global Climate Change.

Cherp, A., Jewell, J., and Goldthau, A. 2011. Governing Global Energy: Systems, Transitions, Complexity. *Global Policy* 2(1), 75–88. https://doi.org/10.1111/j.1758-5899.2010.00059.x.

CIF (Climate Investment Funds). 2018. Energy Access. Available at www.climateinvestmentfunds.org/topics/energy-access.

Climate Tagger. 2018. Renewables Tagger. Available at www.climatetagger.net/climate-tagger-tools/renewabletagger/.

CNP (CarbonNeutral Protocol). 2018. Carbon Neutral. Available at www.carbonneutral.com.

Colgan, J. D., Keohane, R. O., and Van de Graaf, T. 2012. Punctuated Equilibrium in the Energy Regime Complex. *The Review of International Organizations* 7(2), 117–143. https://doi.org/10.1007/s11558-011-9130-9.

Dubash, N. K. and Florini, A. 2011. Mapping Global Energy Governance. *Global Policy* 2, 6–18. https://doi.org/10.1111/j.1758-5899.2011.00119.x.

E4I (Energy 4 Impact). 2018. Energy for Impact: Accelerating Access to Energy. Available at www.energy4impact.org/.

Elliott, D. 2000. Renewable Energy and Sustainable Futures. *Futures* 32(3–4), 261–274. https://doi.org/10.1016/S0016-3287(99)00096-8.

Florini, A. and Sovacool, B. K. 2009. Who Governs Energy? The Challenges Facing Global Energy Governance. *Energy Policy* 37(12), 5239–5248. https://doi.org/10.1016/j.enpol.2009.07.039.

Florini, A. and Sovacool, B. K. 2011. Bridging the Gaps in Global Energy Governance. *Global Governance* 17(1), 57–74. https://doi.org/10.5555/ggov.2011.17.1.57.

Frey, G. W. and Linke, D. M. 2002. Hydropower as a Renewable and Sustainable Energy Resource Meeting Global Energy Challenges in a Reasonable Way. *Energy Policy* 30(14), 1261–1265. https://doi.org/10.1016/S0301-4215(02)00086-1.

Gehring, T. and Oberthür. S. 2009. The Causal Mechanisms of Interaction between International Institutions. *European Journal of International Relations* 15(1), 125–156. https://doi.org/https://doi.org/10.1177/1354066108100055.

Gold Standard. 2018. Our Purpose. Available at www.goldstandard.org/our-story/who-we-are.

GSEP (Global Sustainable Electricity Partnership). 2018. Global Sustainable Electricity Partnership. Available at https://globalelectricity.org.

Heubaum, H. and Biermann, F. 2015. Integrating Global Energy and Climate Governance: The Changing Role of the International Energy Agency. *Energy Policy* 87, 229–239. https://doi.org/10.1016/j.enpol.2015.09.009.

IEA/OECD (International Energy Agency/ Organisation for Economic Co-operation and Development). 2017. *Statistics: CO2 Emissions from Fuel Combust–on - Highlights.* Paris: IEA/OECD.

IRENA (International Renewable Energy Agency). 2015. *REthinking Energy: Renewable Energy and Climate Change.* Abu Dhabi: IRENA.

IRENA (International Renewable Energy Agency). 2018a. *Global Energy Transformation: A Roadmap to 2050.* Abu Dhabi: IRENA.

IRENA (International Renewable Energy Agency). 2018b. About IRENA. Available at irena.org/aboutirena.

IRENA (International Renewable Energy Agency). 2018c. IRENA Coalition for Action About Us. Available at coalition.irena.org/#AboutUs.

IRENA, OECD/IEA, and REN21. 2018. *Renewable Energy Policies in a Time of Transition.* IRENA, OECD/IEA, and REN21.

Karlsson-Vinkhuyzen, S. I. 2010. The United Nations and Global Energy Governance: Past Challenges, Future Choices. *Global Change, Peace & Security* 22(2), 175–195.

Karlsson-Vinkhuyzen, S. I, Jollands, N., and Staudt, L. 2012. Global Governance for Sustainable Energy: The Contribution of a Global Public Goods Approach. *Ecological Economics* 83, 11–18. https://doi.org/10.1016/j.ecolecon.2012.08.009.

Keohane, R. O. and Victor, D. G. 2011. The Regime Complex for Climate Change. *Perspectives on Politics* 9(1), 7–23. https://doi.org/10.1017/S1537592710004068.

Ki-moon, B. 2011. *Sustainable Energy for All: A Vision Statement by Ban Ki-moon, Secretary General of the United Nations.* Washington, DC: SEforALL.

LCTPi (Low Carbon Technology Partnerships initiative). 2018. The Challenge. Available at https://lctpi.wbcsd.org/thechallenge/.

Leal-Arcas, R. and Filis, A. 2013. The Fragmented Governance of the Global Energy Economy: A Legal-Institutional Analysis. *The Journal of World Energy Law & Business* 6(4), 348–405. Available at SSRN: https://ssrn.com/abstract=2296311.

LEDS_GP (Low Emission Development Strategies Global Partnership). 2018 About. Available at ledsgp.org/about/?loclang=en_gb.

Meckling, J. 2018. Governing Renewables: Policy Feedback in a Global Energy Transition. *Environment and Planning C: Politics and Space* 0(0), 1–21.

Newell, P., Phillips, J., and Mulvaney, D. 2011. *Pursuing Clean Energy Equitably Human Development Research Paper 2011/03.* New York: UNDP.

Oberthür, S. 2009. Interplay Management: Enhancing Environmental Policy Integration among International Institutions. *International Environmental Agreements: Politics, Law and Economics* 9(4), 371–391.

Oberthür, S. and Gehring, T. 2006a. *Conceptual Foundations of Institutional Interaction. In Institutional Interaction in Global Environmental Governance: Synergy and Conflict among International and EU Policies.* Cambridge, MA: MIT Press.

Oberthür, S. and Gehring, T. 2006b. Institutional Interaction in Global Environmental Governance: The Case of the Cartagena Protocol and the World Trade Organization. *Global Environmental Politics* 6(2), 1–31.

OECD/IEA (Organisation for Economic Co-operation and Development/ International Energy Agency). 2017. *World Energy Outlook 2017 Executive Summary*. Paris: OECD/IEA.

Pattberg, P., Szulecki, K., Chan, S., and Mert, A. 2009. Assessing the Role and Relevance of the Renewable Energy and Energy Efficiency Partnership in Global Sustainability Governance. In *Enhancing the Effectiveness of Sustainability Partnerships*, edited by D. Vollmer. Washington, DC: U.S. National Academies Press.

Pattberg, P. and Widerberg, O. 2014. *Transnational Multi-Stakeholder Partnerships for Sustainable Development – Building Blocks for Success. IVM Report; No. R-14/31*. Amsterdam: Institute for Environmental Studies.

R20 (R20 – Regions of Climate Action). 2018. Our Mission. Available at https://regions20.org/about-us-2/.

RE100. 2018. RE100. Available at there100.org/re100.

REEEP (Renewable Energy and Energy Efficiency Partnership). 2015. *A New Normal for Green Growth – Annual Report 2014/15*. Vienna: REEEP.

REEEP (Renewable Energy and Energy Efficiency Partnership). 2016a. *Beyond Energy – Annual Report 2016*. Vienna: REEEP.

REEEP (Renewable Energy and Energy Efficiency Partnership). 2016b. *Sustainable Energy Consumption and Production (SECP) in Agriculture and Integrated Waste Management: Research and Training 2016*. Vienna: REEEP.

REEEP (Renewable Energy and Energy Efficiency Partnership). 2017. *Scale and Sustainability – Toward a Public-Private Paradigm in Powering India*. Vienna: REEEP.

REEEP (Renewable Energy and Energy Efficiency Partnership). 2018a. *REEEP Powering Prosperity Beyond the Grid – Annual Report 2018*. Vienna: REEEP.

REEEP (Renewable Energy and Energy Efficiency Partnership). 2018b. About REEEP. Available at www.reeep.org/about-reeep.

REEEP (Renewable Energy and Energy Efficiency Partnership). 2018c. Reegle.Info. Available at www.reeep.org/reegleinfo.

REN21 (Renewable Energy Policy Network for the 21st Century). 2005. Renewables 2005 Global Status Report. Washington, DC: REN21.

REN21 (Renewable Energy Policy Network for the 21st Century). 2015. *REN21 Annual Report 2015: Providing High Quality Information*. Paris: REN21.

REN21 (Renewable Energy Policy Network for the 21st Century). 2016a. *East African Community Renewable Energy and Energy Efficiency Regional Status Report 2016*. Paris: REN21.

REN21 (Renewable Energy Policy Network for the 21st Century). 2016b. *REN21 Annual Report 2–16 – Connecting the Dots: Convening Multi-Stakeholders on Renewable Energy*. Paris: REN21.

REN21 (Renewable Energy Policy Network for the 21st Century). 2017a. *Renewables 2017 Global Status Report*. Paris: REN21.

REN21 (Renewable Energy Policy Network for the 21st Century). 2017b. *UNECE Renewable Energy Status Report 2017*. Paris: REN21.

REN21 (Renewable Energy Policy Network for the 21st Century). 2018a. *REN21 Annual Report 2017*. Paris: REN21.

REN21 (Renewable Energy Policy Network for the 21st Century). 2018b. *Renewables 2018 Global Status Report*. Paris: REN21.

REN21 (Renewable Energy Policy Network for the 21st Century). 2018c About Us. Available at www.ren21.net/about-ren21/about-us/.

Röhrkasten, S. 2015. *Global Governance on Renewable Energy – Contrasting the Ideas of the German and Brazilian Governments*. Wiesbaden: Springer VS.

Röhrkasten, S. and Westphal, K. 2013. *IRENA and Germany's Foreign Renewable Energy Policy – Aiming at Multilevel Governance and an Internationalization of the Energiewende?* Berlin: SWP.

RSB. 2018. Certification. Available at https://rsb.org/certification/.

Sanderink, L. 2019. Governing Renewables Globally: Facing the Frames. Unpublished manuscript.

Sanderink, L., Kristensen, K., Widerberg, O., and Pattberg, P. 2018. *Mapping the Institutional Architecture of Global Energy Governance*. Amsterdam. https://doi.org/10.13140/RG.2.2.22530.38085.

Schäferhoff, M., Campe, S., and Kaan, C. 2009. Transnational Public-Private Partnerships in International Relations: Making Sense of Concepts, Research Frameworks, and Results. *International Studies Review* 11(3), 451–474.

SEforALL (Sustainable Energy for All). 2017. *State of Electricity Access Report 2017*. Washington, DC: SEforALL.

SEforALL (Sustainable Energy for All). 2018a. Mission. Available at www.seforall.org/mission.

SEforALL (Sustainable Energy for All). 2018b. Sustainable Energy for All. Going Further, Faster – Together. 2017 Annual Report. Vienna: SEforALL.

SEforALL (Sustainable Energy for All). 2018c. About Us. Available at www.seforall.org/about-us.

SEforALL (Sustainable Energy for All). 2018d. CEO of SEforALL. Available at www.seforall.org/about-us_ceo-srsg.

Sovacool, B. K. 2013. Expanding Renewable Energy Access with Pro-Poor Public Private Partnerships in the Developing World. *Energy Strategy Reviews* 1(3), 181–192.

Sovacool, B. K. and Florini, A. 2012. Examining the Complications of Global Energy Governance. *Journal of Energy and Natural Resources Law* 30(3), 235–263.

Sovacool, B. K. and Van de Graaf T. 2018. Building or Stumbling blocks? Assessing the Performance of Polycentric Energy and Climate Governance Networks. *Energy Policy* 118, 317–324.

Stokke, O. S. 2001. *The Interplay of International Regimes: Putting Effectiveness Theory to Work. FNI Report 14/2001*. Lysaker: The Fridtjof Nansen Institute.

Suding, P. H. and Lempp, P. 2007. The Multifaceted Institutional Landscape and Processes of International Renewable Energy Policy. International Association for Energy Economics Newsletter Second Quarter: 4–9. Available at www.iaee.org/documents/newsletterarticles/Suding.pdf.

Szulecki, K., Pattberg, P., and Biermann, F. 2011. Explaining Variation in the Effectiveness of Transnational Energy Partnerships. *Governance* 24(4), 713–736.

The Independent Commission on International Development Issues. 1980. *North-South: A Programme for Survival*. London: Pan books.

United Nations. 1998. Kyoto Protocol to the United Nations Framework Convention on Climate Change.

United Nations. 2015. General Assembly Resolution 70/1. Transforming Our World: The 2030 Agenda for Sustainable Development (October 21, 2015).

United Nations 2018. *The Sustainable Development Goals Report 2018*. New York: United Nations. https://doi.org/10.18356/3405d09f-en.

Van de Graaf, T. 2012. Obsolete or Resurgent? The International Energy Agency in a Changing Global Landscape. *Energy Policy* 48, 233–241.

Van de Graaf, T. 2013. *The Politics and Institutions of Global Energy Governance*. Hampshire & New York: Palgrave Macmillan.

WBCSD (World Business Council for Sustainable Development). 2018. Climate & Energy. Available at www.wbcsd.org/Programs/Climate-and-Energy.

WFC (World Futures Council). 2016. *Renewable Energy and Sustainable Development. Accounting for Impacts on the Path to 100% RE*. Hamburg: WFC.

Widerberg, O., Pattberg, P., and Kristensen, K. 2016. *Mapping the Institutional Architecture of Global Climate Change Governance V.2*. Amsterdam: Institute for Environmental Studies.

Wilson, J. D. 2015. Multilateral Organisations and the Limits to International Energy Cooperation. *New Political Economy* 20(1), 85–106.

World Bank. 2015. *Beyond Connections: Energy Access Redefined*. Washington, DC: World Bank.

World Bank and IEA/OECD. 2015. *Sustainable Energy for All. Progress towards Sustainable Energy 2015. Global Tracking Framework*. Washington, DC: World Bank and IEA/OECD.

World Bank and OECD/IEA. 2017. *Sustainable Energy for All. Global Tracking Framework. Progress toward Sustainable Energy 2017*. Washington, DC: World Bank and IEA/OECD.

Zelli, F. 2010. *Conflicts among International Regimes on Environmental Issues. A Theory-Driven Analysis*. PhD Dissertation. Tübingen: Eberhard-Karls University.

Zelli, F. and van Asselt, H. 2010. The Overlap between the UN Climate Regime and the World Trade Organization: Lessons for Post-2012 Climate Governance. In *Global Climate Governance beyond 2012: Architecture, Agency and Adaptation*, edited by F. Biermann, P. Pattberg, and F. Zelli. Cambridge: Cambridge University Press.

Zelli, F., Gupta, A., and van Asselt, H. 2012. Horizontal Institutional Interlinkages. In *Global Environmental Governance Reconsidered*, edited by F. Biermann and P. Pattberg. Cambridge, MA & London: MIT Press.

5

Fossil Fuel Subsidy Reform

Interactions between International Cooperative Institutions. The More, the Merrier?

CLEO VERKUIJL AND HARRO VAN ASSELT

5.1 Introduction

There is increasing recognition that fossil fuel subsidy reform (FFSR) can contribute to a host of environmental, social, and economic objectives, and thereby contribute to achieving both the Sustainable Development Goals (SDGs) and the goals of the Paris Agreement on climate change (e.g. Jakob et al. 2015; Jewell et al. 2018; UNEP 2018). However, at several hundred billion dollars a year (OECD 2018), fossil fuel subsidies persist in both developed and developing economies.

While past research has sought to address this puzzle through the lens of domestic politics – pointing to challenges to reform such as popular opposition, vested interests, interest groups, path dependency, and capacity and data gaps (e.g. Victor 2009; Inchauste and Victor 2017) – international cooperation can also play an important role in promoting, or impeding, FFSR (Smith and Urpelainen 2017; Skovgaard and van Asselt 2018). For instance, while international institutions can adopt new rules, catalyze international commitments, enhance states' accountability, and facilitate information-sharing and capacity-building, there is also a risk that they will struggle to move beyond rhetoric, promote weak, vague, or otherwise inadequate norms, or be perceived to favour certain approaches over others. Where more than one international institution is active at the same time, the door is open for cooperation, as well as competition and conflict between different institutions. This raises questions regarding the institutional coherence of international FFSR governance (see also Chapter 2).

In this chapter, we consider how various international institutions are approaching FFSR governance. First, we briefly introduce the rationale for FFSR (Section 5.2). Next, we discuss the international FFSR governance architecture, with a view to analyzing institutional coherence at the meso level (Section 5.3). This includes the possible emergence of a core norm of FFSR, membership distribution, and the governance functions carried out by the various international institutions active in this area. To further evaluate the degree of coherence in this field, we zoom in on the meso level. Concretely, we examine a subset of three international clubs whose FFSR activities are among the most prominent globally: the Group of 20 (G20), the Asia-Pacific Economic Cooperation (APEC), and the Friends of Fossil Fuel Subsidy Reform. We first introduce the FFSR activities being undertaken by each of these three institutions, and then consider the interlinkages between these activities, as well as efforts to manage them (Sections 5.4–5.5). We conclude by considering implications of our findings for the future management of FFSR governance and the complexity thereof (Section 5.6).

5.2 Fossil Fuel Subsidies: The Rationales for Reform

Fossil fuel subsidies are a form of government support that benefits the producers or consumers of coal, oil, and gas. Both developed and developing countries subsidize fossil fuels: subsidies for consumers are more commonplace in developing countries, while those for producers are found across the board (Bast et al. 2015). Such assistance can come in many guises; some more direct than others. Common types of subsidies include direct transfers of funds; the setting of prices above or below market rates; exceptions or reductions on taxes; favourable loans, loan guarantees, or insurance rates; and preferential government procurement. Support can also be provided in-kind, such as when a government builds infrastructure for the primary or exclusive use of a coal company.

The Organisation for Economic Co-operation and Development (OECD), the International Energy Agency (IEA), and the International Monetary Fund (IMF) have all published estimates of government support to fossil fuel use and consumption. These numbers vary, depending on what valuation method is used; which countries and regions are covered; fluctuations over time; and which definition of a 'subsidy' is being used. However, even by the more conservative OECD and IEA estimates, global fossil fuel subsidies totalled between US $373 and 617 billion per year between 2012 and 2015 (OECD 2018). The IMF's approach, which incorporates the non-priced externalities of fossil fuel production and consumption such as air pollution, traffic congestion, and climate change, suggests the public costs lie much higher: in the range of US $5.3 trillion in 2015 (Coady et al. 2017).

Unsurprisingly, the IMF's broad 'post-tax' interpretation of what constitutes a fossil fuel subsidy has proved controversial. Yet the Fund's approach does help to illustrate the broader societal costs of fossil fuels. Moreover, regardless of the definition used, fossil fuel subsidies are associated with significant economic, social, and environmental impacts. Even excluding non-priced externalities, fossil fuel subsidies can represent a major burden on the public purse, taking up as much as 35 per cent of the public budget in some countries (El-Katiri and Fattouh 2015).

They thereby reduce the investments available for key development sectors such as health and education (Merrill and Chung 2015), representing an important opportunity cost for developing countries in particular. Moreover, while fossil fuel subsidies are often defended as being 'pro-poor', the evidence suggests that such measures tend to be highly regressive, and, perversely, generally benefit those who consume the most energy in society, or powerful interest groups (Arze del Granado and Coady 2012).

But perhaps the most urgent rationale for phasing out fossil fuel subsidies lies in their environmental impact. These subsidies artificially enhance the competitiveness of fossil fuels, potentially locking in unsustainable fossil fuel infrastructure for decades (Asmelash 2016). Indeed, it has been estimated that more than a third of carbon emissions between 1980 and 2010 were driven by fossil fuel subsidies (Stefanski 2014). According to the 2018 Emissions Gap Report by the United Nations Environment Programme (UNEP), phasing out fossil fuel subsidies worldwide could reduce global carbon emissions by up to 10 per cent (UNEP 2018). Government support to fossil fuels also diverts investment from areas such as energy efficiency and renewable energy, while their reinvestment in these areas could bring about important climate change mitigation benefits (Merrill et al. 2016).

Notwithstanding the adverse fiscal, socioeconomic, and environmental effects of fossil fuel subsidies, decades of experience with fossil fuel subsidy reform in various countries attest to the political challenges. While some countries, such as India, Indonesia, and Mexico, have made some progress in reforming subsidies, other countries, such as Nigeria, have struggled to implement or sustain reforms. Reforms have usually been linked to macro-economic factors such as falling fossil fuel prices (Benes et al. 2015) or financial crises, but in many cases domestic political factors, such as the role of special interest groups, a country's institutional and governance capacity, and the political system, play a crucial role in making FFSR a success (Skovgaard and van Asselt 2018).

The importance of macro-economic and domestic political factors in hindering or driving FFSR may suggest that there is no or only a limited role for international cooperation in steering reform. However, as the next section shows, international governance can help drive (or hinder) domestic reform.

5.3 Meso-Level Coherence

5.3.1 Emergence of the FFSR Institutional Complex

As fiscal instruments of energy policy that can have numerous social, economic, and environmental effects, it is not surprising that fossil fuel subsidies are governed by a range of institutions from respective domains, such as energy, trade, and sustainable development (Van de Graaf and van Asselt 2017). However, until well into the previous decade, there were hardly any international institutions focusing specifically on the problems posed by fossil fuel subsidies, or options for their reform.

This changed in 2009, a watershed moment for the international politics of FFSR. Meeting in Pittsburgh in September, G20 leaders made the first international commitment to address fossil fuel subsidies (G20 2009). This commitment was closely followed by a similar pledge by the 21 APEC economies (APEC 2009). Then-US President Barack Obama is widely credited with orchestrating the G20 pledge as he sought to shape his administration's climate legacy, with the economic crisis offering a further window of opportunity to promote new approaches to financial governance (Van de Graaf and Blondeel 2018).

Over the subsequent decade, a range of additional institutions has become active in this field. These include various international organizations such as the OECD, IEA, IMF, and World Bank; additional minilateral coalitions such as the Friends of Fossil Fuel Subsidy Reform (Friends); and nongovernmental organizations (NGOs) such as the Global Subsidies Initiative. The profile of FFSR has further been raised through two further developments: a reference to the need to 'rationalize inefficient fossil fuel subsidies' in SDG target 12.c (UN 2015); and the recent adoption, by twelve members of the World Trade Organization (WTO), of a Ministerial Statement that highlights the need to take this agenda forward in the international trade sphere (IISD 2017).

5.3.2 The Core Norm of Fossil Fuel Subsidy Reform

The starting point of any analysis of a core norm of FFSR is the aforementioned G20 commitment, made at the Group's third leaders' summit in Pittsburgh in September 2009. In their statement, leaders recognized that inefficient fossil fuel subsidies encourage wasteful consumption, distort markets, impede investment in clean energy sources, and undermine efforts to deal with the threat of climate change (G20 2009, preamble). As such, they committed to '[r]ationalize and phase out over the medium term inefficient fossil fuel subsidies that encourage wasteful consumption' (G20 2009, paragraph 29). As mentioned in Section 5.3.1, ministers of APEC adopted a similar pledge later that year (APEC 2009).

While we take this formulation as the general core norm for this chapter, the precise content of the norm of FFSR remains contested and invites different interpretations. First, as mentioned in Section 5.2, there is no universal definition of a 'fossil fuel subsidy', and indeed, different organizations have historically approached this question in various ways (see also Chapter 8). It should be mentioned, however, that a common conception of a fossil fuel subsidy may be increasingly within reach, in particular as official guidance for their measurement has been released in the context of the SDG process (UNEP et al. 2019). Nevertheless, the G20 and APEC commitments leave important qualifiers such as 'inefficient' and 'wasteful consumption' undefined. This has given countries considerable leeway to adopt their own interpretations of their international FFSR commitments (Asmelash 2016; Aldy 2017). As a consequence, countries such as Japan, Saudi Arabia, and the United Kingdom have been able to claim they have no fossil fuel subsidies at all (Van de Graaf and Blondeel 2018). Another issue that remains unclear is by when fossil fuel subsidies need to be phased out. Although a range of stakeholders, including leading investors and insurers (Reuters 2017), have called for a phase-out by 2020, and the G7 adopted a date of 2025, the G20 and APEC commitments do not have a clear timeline (with the G20's reference to the 'medium-term' adding limited guidance).

Despite the textual ambiguity of these initial pledges, the understanding that at least a subset of fossil fuel subsidies ought to be reformed has gained significant global traction over the past decade, including through a reference to FFSR in the UN's 2030 Agenda for Sustainable Development (UN 2015, SDG target 12.c). The European Union has also adopted its own pledge to phase out harmful fossil fuel subsidies by 2020. Indeed, as noted by Rive (2018, 164), the G20 and APEC's initial pledges have been instrumental in 'a reframing of a conception of fossil fuel subsidies as a legitimate government tool to enhance economic development, energy security and welfare into a normative conception that is broadly negative in fiscal and environmental terms'.

Recent developments nevertheless suggest that support for this core norm cannot necessarily be taken for granted. In 2017, the year US President Trump assumed office, G20 leaders omitted FFSR from their declaration for the first time since 2009. While the accompanying G20 Hamburg Climate and Energy Action Plan for Growth (G20 2017) includes a separate FFSR section, the document contains an overall reservation from the United States. Significantly, the 2017 APEC Leader's Declaration also omits a reference to FFSR (APEC 2017).

5.3.3 Membership

The international institutions governing fossil fuel subsidies include government-driven multilateral regimes, international organizations, and clubs involving a small group of countries (see Table 5.1). This diversity notwithstanding, all but one of the institutions outlined in Table 5.1 are public. The only exception is the Geneva-based Global Subsidies Initiative, a programme of the International Institute for Sustainable Development, a Canadian NGO (Lemphers et al. 2018). One explanation for this is that fossil fuel subsidies are, by definition, provided by governments, who therefore have a key role in addressing them. The dominant role of public institutions suggests that the global governance architecture on FFSR is somewhat less

Table 5.1 *Overview of governance functions across different types of institutions (public, hybrid, private) for fossil fuel subsidy reform.*

	Public	Hybrid	Private
Standards & Commitments	G7, UNFCCC, UN (SDGs)		
Operational			
Information & Networking	IEA, OECD, OPEC, WTO, UNEP		
Financing			
Standards & Commitments; Operational			
Operational; Information & Networking			Global Subsidies Initiative
Information & Networking; Financing	IMF, World Bank		
Standards & Commitments; Information & Networking	APEC, Friends of Fossil Fuel Subsidy Reform, G20		
Standards & Commitments; Financing			
Operational; Financing			

institutionally complex than that of other subfields in the climate-energy nexus that heavily involve civil society and the private sector (see Chapter 3).

5.3.4 Governance Functions

The range of international institutions working on FFSR covers the full gamut of governance functions. The diversity of these institutions engaged is notable: they range from those whose core mandates concern fiscal governance (e.g. the G20, OECD, IMF, and World Bank), to trade liberalization (WTO and APEC), to energy (IEA and OPEC), to the environment and climate change (UNEP and the UN Framework Convention on Climate Change (UNFCCC)).

In terms of standards and commitments, forums such as the G20, G7, APEC, and Friends have made pledges or otherwise publicly promoted the norm of FFSR.

The UN's 2030 Agenda, through SDG 12.c., also encourages countries to 'rationalize' inefficient fossil fuel subsidies. Fossil fuel subsidy review processes, such as those organized by the G20 and APEC, provide an opportunity for individual countries to pledge to address certain subsidies; while the UNFCCC, through its system of nationally determined contributions (NDCs), allows countries to make similar national commitments.

Regarding information and networking, organizations such as the OECD, IEA, IMF, World Bank, Friends, and Global Subsidies Initiative conduct research to clarify the scale of subsidies provided. At the same time, forums such as the G20, APEC, and the WTO create mechanisms for countries to report on their fossil fuel subsidies, although notification rates are patchy. The UN Environment Programme supports countries in better understanding the extent of their fossil fuel subsidies through the development of international indicators. As discussed in more detail in Section 5.4, the Friends engage in behind-the-scenes networking to further promote international reform efforts.

Operational activities such as technical assistance and capacity-building are provided by international organizations such as the World Bank, Friends, and Global Subsidies Initiative, including through publications, events, and online webinars.

Finally, organizations such as the World Bank and Friends have made financing available to help developing countries undertake reform, while structural adjustment policies implemented by the IMF and the World Bank have at times involved FFSR.

5.3.5 Summary: Coherence at the Meso Level

The coherence of the institutional complex for fossil fuel subsidies may seem limited at first sight. There is not a single definition that all institutions adhere to and, as Section 5.2 pointed out, existing definitions and ways of measuring subsidies differ widely. Moreover, international institutions seem to address fossil fuel subsidies for very different reasons, from fiscal to environmental. Lastly, membership is heavily skewed toward public institutions, and the role of private and hybrid institutions is limited.

However, the level of inconsistency should not be exaggerated. First, notwithstanding divergences in the way fossil fuel subsidies are defined and measured, there are also important similarities in the various definitions published by the OECD, IEA, and IMF (Koplow 2018), and joint estimates by the OECD and IEA have been published (OECD 2018). Second, a deeper dive into the types of governance functions fulfilled by various institutions suggests that there is a certain synergy emerging, with all governance functions being fulfilled by several institutions (Table 5.1). While some forums have been instrumental in agenda-setting, and the formulation of broad commitments (e.g. the G20, APEC, Friends), other organizations have focused on providing information on subsidies and their impacts (e.g. the OECD, IEA, IMF), while yet others have been key in supporting FFSR on the ground through, for instance, financing and capacity-building (e.g. World Bank, Global Subsidies Initiative).

Partially, this distribution is the result of active coordination. As a notable example, the G20's 2009 commitment to phase out inefficient fossil fuel subsidies largely preceded the availability of robust data into global and domestic fossil fuel subsidies (Van de Graaf and Blondeel 2018). However, in their Pittsburgh Statement, G20 leaders requested 'relevant institutions', such as the IEA, the Organization of the Petroleum Exporting Countries (OPEC), the OECD, and the World Bank (the 'IGO-4'), to provide an analysis of the scope of energy subsidies and suggestions for the implementation of the G20's reform efforts (G20 2009), with several reports issued thus far, including on how such reforms can be made while assisting the poor (IEA, OECD, and World Bank 2010a, 2010b, 2011; World Bank 2014).

In terms of the terminology introduced in Chapter 2, our overall assessment of the meso-level coherence therefore falls between synergy and division of labour.

5.4 Micro-Level Coherence

The remainder of this chapter will focus on three forums in particular to scrutinize coherence at the micro level of the FFSR subfield. These forums are: the G20, APEC, and Friends. Despite differing overarching mandates and approaches, they have been among the most active in this subfield, including through proactive promotion of an international norm on FFSR. All three are clubs involving a limited number of economies, who moreover began to address fossil fuel subsidies roughly around the same time (2009–2010). As such, their activities in this space are comparable. Nevertheless, as will be discussed, there are differences in their approaches as well. To shed light on how international FFSR governance is impacted by the parallel efforts of these three forums, we consider how each is addressing FFSR, to what extent their approaches are consistent, and the management of the interaction between them.

5.4.1 Institutions under Scrutiny

5.4.1.1 The Group of 20

The Group of 20 was established in 1999 in response to the Asian financial crisis. During the financial crisis of 2008, the Group's status was elevated to that of a leaders' summit, with members' heads of state and government convening once or twice a year since to address issues relating to global economic governance and reform (Wade 2011). Comprising nineteen of the world's largest developed and developing economies,[1] as well as the European Union, the Group accounts for some two-thirds of global population; 85 per cent of global gross domestic product (GDP) (Kim and Chung 2012), and 75 per cent of global greenhouse gas emissions (Climate Transparency 2015).

While economic governance is the Group's *raison d'être*, climate change has featured on the leaders' agenda from the beginning (Kirton and Kokotsis 2015). Assessments of its performance in this regard have often been cautiously positive (Van de Graaf and Westphal 2011; Garnaut 2014; Kirton and Kokotsis 2015), with commentators identifying the Group's flexibility over topics and time; its ability to exploit issue linkages; and 'a sense of being equal' among members (Kim and Chung 2012) as advantages. However, important drawbacks of the 'exclusive minilateralism' pursued by the G20 have also been identified, including a lack of legitimacy, transparency, and accountability, in particular when compared to bodies with a more universal membership, such as the UNFCCC (Eckersley 2012; Kim and Chung 2012).

As mentioned earlier, the G20's 2009 FFSR pledge has been instrumental in elevating the issue to the international agenda. In addition to this role, leaders at the 2009 Pittsburgh summit also agreed to prepare and report on implementation strategies and time frames for the rationalization and phase-out of inefficient fossil fuel subsidies (G20 2009). To facilitate this work, the G20 established a working group on energy in which energy experts, under supervision of Finance and Energy ministers, reviewed the fossil fuel subsidies in their countries (Kim and Chung 2012). However, despite its potential to enhance transparency in the area of fossil fuel subsidies, the results of this exercise have been described as 'meagre' (Van de Graaf and Westphal 2011, 28) and 'disappointing' (de Jong and Wouters 2014, 34), with almost half of the G20 countries providing little or no further information on their subsidies (Van de Graaf and Blondeel 2018).

A more in-depth process to increase the transparency on subsidies of a subset of G20 countries is currently ongoing. This goes back to June 2012, when G20 Leaders requested Finance ministers to explore options for a voluntary peer review (VPR) process (G20 2012). In February of the following year, G20 Finance Ministers committed to undertake such a process, and, several months later, released a corresponding methodology (G20 Energy Sustainability Working Group 2013). Reciprocal VPRs have been conducted by China and the United

[1] Argentina, Australia, Brazil, Canada, China, France, Germany, India, Indonesia, Italy, Japan, South Korea, Mexico, Russia, Saudi Arabia, South Africa, Turkey, the United Kingdom, and the United States.

States, as well as Germany and Mexico. Reviews between Indonesia and Italy, and Argentina and Canada, have also been announced. While the VPR process appears to provide opportunities for domestic and bilateral learning in the area of FFSR,[2] engagement in the voluntary process does not necessarily guarantee enhanced transparency. Germany's VPR in particular has been accused of 'ignoring the majority of fossil fuel subsidies' in the country (ODI 2017).

5.4.1.2 The Asia-Pacific Economic Cooperation

The Asia-Pacific Economic Cooperation was created in 1989 as a regional forum in response to the increasing economic interdependence of the region (Elek 2005). Consisting of twenty-one developed and developing countries in the Asia-Pacific region, APEC countries account for half of global trade, 60 per cent of world GDP (APEC 2018), and 60 per cent of global energy demand (IEA 2017). Consequently, energy policy developments within APEC have significant impacts on global energy trends (IEA 2017).

Although action on climate change and energy has not been a central focus of APEC's activities, since 2009 the forum has engaged in a range of activities related to FFSR. Just weeks after the G20's 2009 commitment in Pittsburgh, APEC leaders similarly pledged to 'rationalise and phase out over the medium term fossil fuel subsidies that encourage wasteful consumption', while 'recognising the importance of providing those in need with essential energy services' (APEC 2009).[3] As with the G20, enhancing transparency of existing subsidies was a first step in this effort. Meeting in Japan in 2010, APEC Energy Ministers instructed the group's Energy Working Group (EWG) to provide an initial assessment of fossil fuel subsidies in the region (APEC Energy Ministers 2010). In 2011, APEC Leaders meeting in Honolulu agreed to set up a 'voluntary reporting mechanism' that allows members to self-report progress toward reform (APEC 2011).

Building on their peer review experiences in the areas of renewable energy and energy efficiency,[4] APEC economies have also engaged in their own VPR process for fossil fuel subsidies.[5] Guidelines for VPRs were adopted in November 2013 (APEC EWG 2013), with Peru as the first APEC economy to undergo review.[6] Additional VPRs have been conducted for New Zealand (2015), the Philippines (2015), Chinese Taipei (2016), and Vietnam (IEA 2017). The APEC VPR process reviews fossil fuel subsidies in the volunteer economy, facilitated by the EWG and

[2] Interview with Ronald Steenblik, former Senior Trade Policy Analyst, OECD, 20 July 2018.
[3] Like the G20, APEC Leaders have reaffirmed this commitment in subsequent years (e.g. APEC 2010, 2011, 2012).
[4] Interview with Phyllis Genther Yoshida, former Lead Shepherd, Energy Working Group, APEC, 19 July 2018.
[5] Like APEC, we employ the terminology 'economy' or 'member' rather than 'country' given that APEC members Chinese Taipei and Hong Kong are not typically treated as independent sovereign entities.
[6] Interview with Ananth Chikkatur, Manager, ICF International, 20 July 2018.

FFSR Secretariat, which was established to assist developing economies through coordination of peer review activities and provision of technical and logistical support.[7] The results of the reviews, including policy reform recommendations, are shared to help disseminate lessons learned and best strategies for reform (IEA 2017).

Compared to the G20, APEC includes more developing countries.[8] Perhaps reflecting this more diverse membership, capacity-building has also been an important focus of APEC's FFSR work (APEC 2013), with dedicated FFSR capacity-building workshops held in Honolulu (APEC EWG 2015) and Jakarta (APEC EWG 2017).

5.4.1.3 The Friends of Fossil Fuel Subsidy Reform

The Friends is an informal coalition of countries set up in June 2010 'to build political consensus on the importance of fossil fuel subsidy reform' (GSI n.d.). Currently comprising nine states – Costa Rica, Denmark, Ethiopia, Finland, New Zealand, Norway, Sweden, Switzerland, and Uruguay – the group's establishment was directly inspired by the G20's 2009 commitment to phase out fossil fuel subsidies (Rive 2016). Indeed, the Friends explicitly identifies itself in relation to, and in contrast with, the G20: as 'an informal group of non-G20 countries' (Friends, n.d.).[9] Informal coalitions of 'Friends' that bring together countries with similar views around particular issue areas have become a familiar phenomenon in international affairs, including in the areas of trade, development, environment, and disarmament (Rive 2018). The Friends has elicited comparisons with the 'Friends of Fish': a group of developed and developing countries working within the WTO to promote sustainable fishing practices and elimination of harmful fishing subsidies (Young 2017).

The Friends was established by New Zealand, which continues to play a leading role (Rive 2016). The Global Subsidies Initiative performs a support function for the group (Lemphers et al. 2018). Coordination takes place at the sidelines of biannual meetings of the OECD Joint Working Party on Trade and Environment and other international meetings, through issue-specific meetings of technical experts, and through monthly conference calls.[10]

The Friends has primarily engaged in 'soft' activities in its efforts to promote FFSR. This is in line with Rive's (2018, 158) observation that 'the effectiveness of Friends groups on international norm and policy development and negotiations

[7] Ibid.

[8] These include Chile, Papua New Guinea, Peru, the Philippines, Thailand, and Vietnam.

[9] It is worth noting, however, that Denmark, Finland, and Sweden are all members of the EU, which in turn is a G20 member.

[10] Interview with senior official, Ministry of Foreign Affairs and Trade, Government of New Zealand, 9 August 2018.

largely does not depend on securing and wielding political "hard" power ... [i]nstead, it depends on their ability to network, influence, innovate, problem solve and profile raise'.

In this regard, one key forum that the Friends has focused on is the WTO. Through informal lobbying efforts as well as public outreach activities spearheaded by New Zealand, the group has been instrumental in the adoption, at the 11th WTO Ministerial Conference, of a Ministerial Statement on fossil fuel subsidies. In the statement, twelve signatories urge the WTO to advance the discussion on fossil fuel subsidies, and request transparency and reform of inefficient fossil fuel subsidies that encourage wasteful consumption. Such engagement has the potential to make a significant dent in fossil fuel subsidies globally, given the WTO's previous experience with subsidies reform, for instance in the area of agriculture (Verkuijl et al. 2019). Besides the international trade space, the Friends have also promoted FFSR within the UNFCCC process, advocating, among others, the inclusion of reform plans by countries in their NDCs (Merrill et al. 2016).

Part of the Friends' approach also appears to lean on 'leading by example', with three Friends' Members having undergone self- or peer reviews of their fossil fuel subsidies (Finland and Sweden have conducted independent reviews, while New Zealand's was completed under the auspices of APEC). However, it is worth noting that while at least thirteen countries have made reference to fossil fuel subsidies in their NDCs to date, this includes only two Friends countries: Ethiopia and Costa Rica (Terton et al. 2015; Merrill et al. 2016).

Another key output of the Friends was the release of a Fossil Fuel Subsidy Reform Communiqué in April 2015, which invites all countries, companies, and civil society organizations to join in supporting accelerated action to eliminate inefficient fossil fuel subsidies (Friends 2015). Although it remains to be seen to what extent endorsement will lead to meaningful stakeholder engagement, the document has broadened the range of actors overtly committed to the cause of FFSR. These now include twenty-eight non-G20/APEC countries as well as a host of international organizations and NGOs, and associations representing more than 90,000 investors and corporations (IISD 2016; Rive 2018).

Since the group's inception, developed country Friends members have also contributed to reform efforts through financial support to the FFSR-related activities of organizations such as the World Bank, IMF, OECD, and the Global Subsidies Initiative[11] (see Merrill et al. 2016). Finally, the Friends also contribute to capacity-building through side events at international meetings and webinars on FFSR (Friends 2018).[12]

[11] Interview with former senior official, Ministry of Foreign Affairs and Trade, Government of New Zealand, 19 September 2018.

[12] Ibid.

5.4.2 Interlinkages

The activities of the G20, APEC, and Friends intersect in various ways, even as the three groups show variations in terms of membership and geographic scope, the norms they promote, and their governance functions. Taking these dimensions as a starting point, this section examines the coherence between the three groups and their FFSR activities.

There appears to be general consistency with respect to the core norm that the G20, APEC, and Friends espouse in relation to FFSR. In their respective Leaders' Declarations, both the G20 and APEC explicitly commit to 'rationalise and phase out' fossil fuel subsidies that are 'inefficient'. The G20 further qualifies its reform commitment by singling out subsidies that 'encourage wasteful consumption'. On the other hand, the Group's pledge is strengthened by the inclusion of a 'medium-term' timeline for reform. Although APEC's initial pledge in 2009 contained an identical reference, this was dropped in some of the later iterations (e.g. APEC 2010, 2011, 2012). The Friends' efforts in this area appear geared to 'help remind countries' to keep this topic on their agendas through various diplomatic channels.[13] Notably, however, the norm espoused in the Friends' 2015 Communiqué does not meaningfully depart from those of other previous pledges.[14] Similar to the G20 and APEC commitments, the document omits a reference to a specific date for achievement of FFSR and leaves the concept of a subsidy undefined.

As noted, the textual ambiguity of the G20 and APEC's pledges has enabled several economies in these groups to avoid taking any meaningful action on FFSR. While it is unsurprising that certain group members – particularly those with large subsidies or a big fossil fuel industry – may seek to maintain the opacity of the FFSR norm, it is less evident why a group such as the Friends has not succeeded in further crystallizing it. One explanation may lie in the fact that the Friends has trod a fine line between trying to enhance ambition while simultaneously seeking not to 'alienate' other governments,[15] particularly those from the G20 and APEC.[16] Another reason relates to a possible strategic value in maintaining the ambiguity of the term 'subsidy'. A flexible definition can be conducive to reform as it enables

[13] Interview with Laura Merrill, Manager, Global Subsidies Initiative, and Senior Policy Advisor, International Institute for Sustainable Development, 10 May 2017.

[14] Rive (2018, 164) suggests the Communiqué's norm might be stronger than the G20's and APEC's pledges given its reference to the 'elimination' of fossil fuel subsidies. Practically speaking, however, there is little difference between the concepts of 'elimination' and 'phasing out', particularly since the phrasing of the Communiqué is not a commitment as such, but rather highlights reasons for why FFSR is needed.

[15] Interview with senior official, Ministry of Foreign Affairs and Trade, Government of New Zealand, 9 August 2018.

[16] Documents obtained from the New Zealand Ministry of Foreign Affairs and Trade highlight a 'risk' that 'G20 members may take a less favourable view of New Zealand/Friends advocacy on this issue . . . notwithstanding that the primary function of the Friends group would be to encourage implementation of commitments *already made* by the G20' (New Zealand Ministry of Foreign Affairs and Trade 2010, 2, emphasis in original).

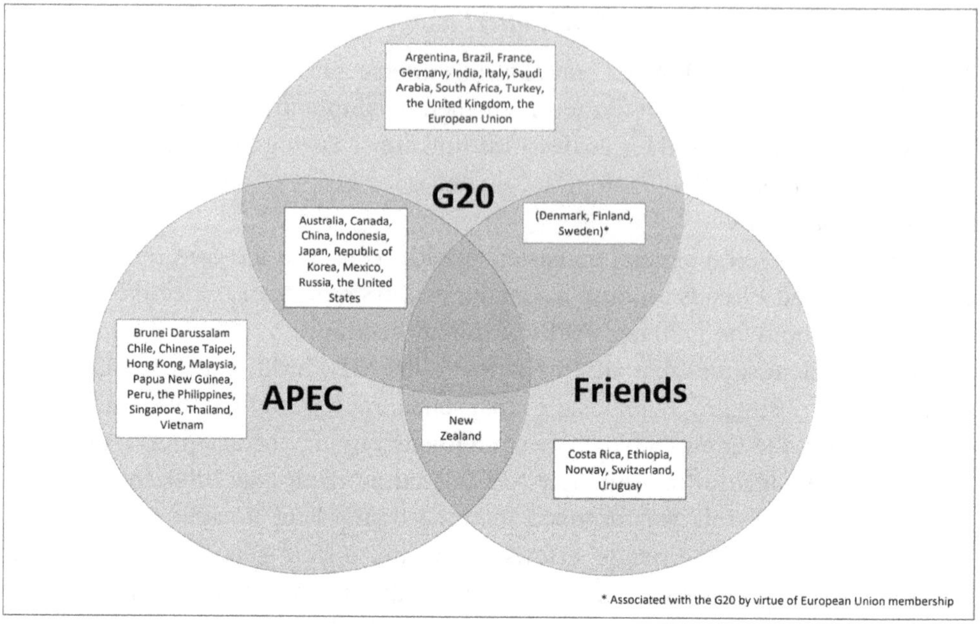

Figure 5.1 Membership of the G20, APEC, and Friends.

governments to engage at a level they feel comfortable with, rather than setting 'too high a threshold' for action.[17]

As regards membership, Figure 5.1 illustrates the degree of overlap between members of the G20, APEC, and Friends. As shown, the majority of G20, APEC, and Friends economies (28 out of 41) only belong to one of these three groups. However, all three groups display at least some degree of overlapping membership, with thirteen members associated with two of the groups. Since countries are unlikely to pursue duplicative or contradictory policies in different forums with similar goals, we can hypothesize that such overlaps will lead to more coherent governance. We can further expect coherence to be enhanced where membership overlaps strongly and where pivotal actors with the ability to influence dynamics straddle multiple groups (Gehring and Faude 2014).

Against this backdrop, it is notable that nine economies are a member of both the G20 and APEC. Moreover, two key global players – the United States and China – have been advocates of FFSR across both groups,[18] including by engaging in some of the first peer reviews, and, in the case of the United States, giving

[17] Interview with senior official, Ministry of Foreign Affairs and Trade, Government of New Zealand, 9 August 2018; and Interview with former senior official, Ministry of Foreign Affairs and Trade, Government of New Zealand, 19 September 2018.

[18] Interview with Ronald Steenblik, former Senior Trade Policy Analyst, OECD, 20 July 2018.

impetus to the initial FFSR pledges made in these forums. Yet, although these interlinkages may increase the likelihood of a common approach between both groups, important differences between the G20 and APEC remain in terms of both their membership and mandate. As such, the mere existence of the G20's 2009 FFSR pledge did not make such a commitment a *fait accompli* in APEC: APEC economies' own interest in fiscal reform at the time was the decisive factor.[19]

Beyond G20-APEC overlaps, New Zealand's membership of both APEC and the Friends also stands out. As one of the strongest proponents of FFSR in APEC, the country has been 'keen to lead by example' in the group, for instance by being among the first to undergo peer review and by seeking to set a 'good benchmark' in doing so.[20] Seeing their APEC activities as 'part of their Friends work', the country has moreover sought to work with like-minded APEC members to inspire APEC to take up similar commitments to the G20.[21] As with China and the United States' overlapping memberships, it is likely that New Zealand's membership of both groups has strengthened the consistency and complementarity between the Friends and APEC's approaches, including by allowing other APEC members to draw on New Zealand's expertise in this area.[22] By contrast, there is little evidence that G20 dynamics have been affected by Denmark, Finland, and Sweden's association with both the Friends and – by virtue of their European Union membership – the G20: presumably a result of competing visions on the topic of reform among the Union's Member States.

While overlapping membership has thus helped to enhance consistency between the three groups, the fact that members *diverge*, particularly in terms of their geographic scope, also contributes to the complementarity of their actions. Indeed, by following on the G20's FFSR pledge, APEC committed eleven additional non-G20 economies in the Asia-Pacific region to FFSR.[23] Similarly, the Friends' 2015 Communiqué broadened support for FFSR to an additional twenty-eight countries and numerous other stakeholders. The three groups' FFSR review activities have also been complementary in scope: at the time of writing, twelve economies had undergone, or committed to undertake, peer reviews under APEC and the G20, with an additional two Friends members having completed self-reviews.

Lastly, in terms of governance functions, there is a significant degree of overlap between the FFSR activities of the G20 and APEC, with both forums engaging in standard and commitment setting, as well as information and networking activities through their progress tracking and peer reviews. Compared to the G20, APEC

19 Interview with Phyllis Genther Yoshida, former Lead Shepherd, Energy Working Group, APEC, 19 July 2018.
20 Interview with senior official, Ministry of Foreign Affairs and Trade, Government of New Zealand, 9 August 2018.
21 Interview with Phyllis Genther Yoshida, former Lead Shepherd, Energy Working Group, APEC, 19 July 2018.
22 Ibid.
23 This includes Chinese Taipei and Hong Kong.

appears to place a stronger emphasis on building the capacity of its members to engage in FFSR by facilitating operational activities such as capacity-building. Nevertheless, these differences should not be overstated. As observed by Steenblik,[24] peer review – while primarily related to information – can also be regarded as an important means of promoting capacity-building, allowing developing and developed countries alike to create a better understanding of the types of fossil fuel subsidies that exist, and ways of addressing them.

There are nonetheless nuances in the way the G20 and APEC have approached their activities. Although much of the G20's FFSR work has been concentrated in its energy working group, its original pledge was coordinated by finance ministers, who have remained heavily involved in this topic.[25] APEC's reform activities, on the other hand, have largely been restricted to the forum's EWG and energy ministers: a deliberate choice on behalf of the forum's FFSR proponents, who feared that involvement of senior finance officials would have rendered this work 'too political'.[26] By allowing APEC to draw on its EWG's experience in delivering on projects in other areas, including peer reviews on renewable energy and energy efficiency, this approach enabled APEC economies to complete the first fossil fuel subsidy review as early as July 2015. By contrast, the G20's approach to peer reviews has been more political, including through the 'pairing' of a developed and developing country review in every review cycle, resulting in a more drawn-out process.[27]

Both groups' approaches to VPRs may be associated with certain advantages. By enabling individual countries to undergo peer review once they are ready, APEC's approach has allowed for a quicker succession of reviews than the G20 approach, which is based on willing pairs of countries stepping forward. On the other hand, it is notable that all the G20 members of APEC have conducted their peer reviews under the auspices of the G20, which is perhaps associated in the public's mind with greater political prestige.[28] One challenge for both groups is how to maintain momentum for VPRs going forward. Naturally, those countries most eager to undergo review were among the first to volunteer, while some remaining countries are more reluctant to engage: for example, several maintain they have no inefficient subsidies in the first place, or want to delay committing to a review for reasons of political timing.[29] Moreover, a backlog of peer reviews has

[24] Interview with Ronald Steenblik, former Senior Trade Policy Analyst, OECD, 20 July 2018.
[25] Interview with Phyllis Genther Yoshida, former Lead Shepherd, Energy Working Group, APEC, 19 July 2018.
[26] Ibid.
[27] Interview with Ronald Steenblik, former Senior Trade Policy Analyst, OECD, 20 July 2018; and Interview with Phyllis Genther Yoshida, former Lead Shepherd, Energy Working Group, APEC, 19 July 2018.
[28] Interview with Ronald Steenblik, former Senior Trade Policy Analyst, OECD, 20 July 2018.
[29] Ibid.

reportedly accumulated under APEC, as funding for these efforts, which had previously come largely from the United States, has not been renewed.[30]

The Friends' activities have some overlaps with those of the G20 and APEC. Much like the G20 and APEC's reform pledges, the Friends' 2015 Communiqué focuses on setting standards and commitments for FFSR. Similarly, some members have engaged in information and networking by undergoing fossil fuel subsidy reviews. However, in line with the Friends' consensus-building role, their activities have generally been more externally focused than those of the other two groups. While the G20 and APEC's work largely revolves around their member base, the Friends have actively engaged in operational activities such as events and webinars to influence third actors, including G20 and APEC members, as well as those involved in forums such as the WTO and UNFCCC. Key achievements in this regard include socialization of the concept of peer review within the G20 and APEC,[31] and the adoption of the Ministerial Statement on FFSR at the WTO. Individual Friends members have also provided financing for FFSR through their aid budgets. By seeking to strengthen existing reform efforts and spread such efforts to new forums, the Friends' activities seem to provide a useful complement to the G20 and APEC's internal efforts.

5.4.3 Mechanisms

All three consistency mechanisms identified in Chapter 2 are reflected in the dynamics between the G20, APEC, and Friends.

Normative mechanisms, whereby *the norms and rules of one institution impact on those of another*, seem to be at play with regard to all three groups' public FFSR announcements. After the G20 announced its FFSR commitment in September 2009, APEC quickly followed suit with an almost identical pledge to 'rationalise and phase out' such subsidies, while leaving important questions about the scope of these measures and end-date for their phase-out, unaddressed. Although the Friends' Communiqué does not represent a direct commitment, the document similarly mirrors the key facets of the G20 and APEC pledges, even where those fall short on ambition and clarity. From a long-term perspective, there may be value in such a prudent approach, as 'speaking the same language' arguably allows for more possibilities for the Friends to engage with the other two groups, including their more hesitant members.

[30] Ibid.

[31] Interview with senior official, Ministry of Foreign Affairs and Trade, Government of New Zealand, 9 August 2018.

Cognitive mechanisms, whereby *knowledge and information are shared across institutions*, are similarly present across all three groups. Like many other organizations working on energy, the APEC Secretariat is invited to provide brief oral reports on its activities at G20 EWG meetings.[32] Since the G20 lacks a formal secretariat, the Group is not offered a similar platform within APEC, although information exchange is facilitated by the two groups' overlapping memberships.[33] Indeed, the APEC EWG guidelines encourage APEC members undergoing a VPR through the G20 to share the results with members of the APEC EWG 'in order to transfer lessons learned from that process to all Members' (APEC EWG 2013, 6).[34] To support mutual learning, APEC's capacity-building workshops have also featured talks on G20 experiences (APEC EWG 2015, 2017).[35]

Similarly, the Friends have occasionally held observer status in G20 meetings[36] while 'invited guests' from non-Friends countries and international organizations have also participated in the Friends' six-monthly meetings.[37] Friends' side events on the margins of meetings of the UNFCCC and the World Bank and IMF have also seen the involvement of representatives from G20 and/or APEC countries such as India, Indonesia, the Philippines, and Mexico (Friends 2017; Sanchez 2017). In 2013, New Zealand's then-Ambassador to the EU and the North Atlantic Treaty Organization (NATO) was invited to present on peer review at a workshop hosted by the Russian G20 Presidency.[38]

Behavioural mechanisms are also present in how the three groups interact, with various examples of *institutions being impacted by the functional and strategic behaviour of their members and other actors*. Indeed, as discussed in more detail in Section 5.5, such mechanisms go to the heart of the Friends' activities, which are directed at monitoring and influencing the FFSR activities of the other two groups, including through lobbying and capacity-building activities. These behavioural dynamics are also a product of interlocking memberships. Countries such as the United States and China have helped to push reform commitments and peer review efforts forward in both the G20 and APEC, while New Zealand has actively sought to promote the FFSR agenda through its APEC membership.

[32] Interview with Ronald Steenblik, former Senior Trade Policy Analyst, OECD, 20 July 2018.
[33] Ibid.
[34] These guidelines further note that 'any efforts undertaken in APEC should be complementary to, and not duplicative of, ongoing efforts in the G20' (APEC EWG 2013, 1), and that '[t]he APEC [review process] was closely coordinated with the ongoing efforts in the G20, similar to the voluntary reporting mechanism . . . so that it will be complementary and not duplicative of G20 efforts' (APEC EWG 2013, 1–2).
[35] Interview with Ronald Steenblik, former Senior Trade Policy Analyst, OECD, 20 July 2018.
[36] Interview with Laura Merrill, Manager, Global Subsidies Initiative, and Senior Policy Advisor, International Institute for Sustainable Development, 10 May 2017.
[37] Interview with senior official, Ministry of Foreign Affairs and Trade, Government of New Zealand, 9 August 2018.
[38] Ibid.

5.5 Micro-Level Management

As noted, there is a significant level of consistency between the G20, APEC, and Friends with regard to the core norm they promote. Although their approaches vary to some extent, the three groups also fulfil governance functions in a synergistic fashion. Their activities further complement each other in terms of their geographical scope. As we discussed earlier, this relatively symbiotic relationship may in part be explained by the groups' overlapping memberships.

In addition to such overlaps, the different groups and their members also engage in formal and informal coordination efforts. This includes creating spaces for attending one another's meetings and external events and, in the case of APEC, institutional encouragement to its members to share experiences from the G20 peer review process with other members.

Most interaction management between the three groups seems to take place informally, however, including through meetings on the sidelines of international events;[39] outreach to Friends members for their expert knowledge;[40] and outreach of Friends members to other countries, particularly in advance of G20 summits.[41] The Friends' Communiqué was furthermore drafted with the involvement of both the United States and France as well as the IEA, IMF, and OECD.[42]

Even where interaction has not been direct, it is furthermore clear that the G20, APEC, and Friends have kept abreast of one another's FFSR activities. As described earlier, APEC economies were well aware of the way the G20 was approaching its FFSR pledge and its VPRs, which helped inform its decision to follow a slightly different tack. Furthermore, Yoshida notes that APEC took care to ensure that the peer reviews undertaken under APEC did not undermine momentum in the G20 by co-opting members from this Group.[43] Meanwhile, APEC developments were also tracked by G20 members. For instance, observing that progress under the G20 was less forthcoming than under APEC, the United States volunteered to undergo a VPR under the former, rather than the latter.[44]

Through what can be characterized as a 'broker' role, the Friends have also proactively sought to enhance the complementarity of the three groups' actions. For instance, developments at the G20 provided New Zealand and others with leverage to ensure similar efforts were undertaken under APEC, both in terms of

[39] Ibid.
[40] Interview with Phyllis Genther Yoshida, former Lead Shepherd, Energy Working Group, APEC, 19 July 2018.
[41] Interview with Laura Merrill, Manager, Global Subsidies Initiative, and Senior Policy Advisor, International Institute for Sustainable Development, 10 May 2017; and Interview with senior official, Ministry of Foreign Affairs and Trade, Government of New Zealand, 9 August 2018.
[42] Interview with Laura Merrill, Manager, Global Subsidies Initiative, and Senior Policy Advisor, International Institute for Sustainable Development, 10 May 2017.
[43] Interview with Phyllis Genther Yoshida, former Lead Shepherd, Energy Working Group, APEC, 19 July 2018.
[44] Ibid.

the adoption of reform commitments as well as the introduction of a peer review process in this area.[45] In undergoing its peer review, New Zealand also involved representatives from China, which in turn informed China's understanding of what the review could look like in a G20 context.[46]

5.6 Conclusions

Over the past decade, more than a dozen international energy, economic, environmental, and trade institutions have sought to promote FFSR in different ways. They have done so by providing financial and other incentives to implement reform, coercing states to undertake reform, diffusing the emerging norm of FFSR, and disseminating information about fossil fuel subsidies and their adverse impacts. However, little is still known about how the efforts of these various institutions 'add up', and the extent to which the activities in different institutions complement or contradict each other. This chapter has sought to shed light on this question by assessing the coherence of the institutional complex governing FFSR.

At the meso level, we identify an emerging division of labour in this subfield. Institutions such as the G20, APEC, and Friends play an important role in setting agendas and commitments (and to a lesser extent, sharing information), while organizations such as the OECD, IEA, and IMF engage in information-sharing on the scale of subsidies and their impact. In parallel, organizations such as the World Bank and Global Subsidies Initiative have emphasized operational activities for capacity-building and implementation purposes. Where activities do overlap, they generally appear to reinforce one another.

Zooming in on a subset of FFSR actors, the activities of the G20, APEC, and Friends have been among the most prominent in the field. Together, these three groups cover forty-one economies and a range of activities from standard and commitment setting, information and networking, operational activities, to financing. We find their efforts in this regard to be consistent with one another, and in many cases complementary. For instance, while their membership partially overlaps, the G20 and APEC's peer review activities target different countries, thereby expanding the geographic reach of such efforts. In addition, many of the efforts undertaken by the Friends and their members have been intentionally directed toward enhancing reform efforts under the G20 and APEC. This high level of consistency appears to be the result of planned coordination between institutions

[45] Interview with senior official, Ministry of Foreign Affairs and Trade, Government of New Zealand, 9 August 2018.

[46] Interview with senior official, Ministry of Foreign Affairs and Trade, Government of New Zealand, 9 August 2018; and Interview with Ronald Steenblik, former Senior Trade Policy Analyst, OECD, 20 July 2018.

and overlapping memberships, as well as a proactive brokering role taken on by some countries, including the Friends.

Despite these synergies, country-level progress on reform remains limited. Public funding for fossil fuel consumption and production continues to total many billions of dollars each year, including in G20 and APEC economies (Bast et al. 2015; Rentschler and Bazilian 2017). While the many domestic political factors impeding FFSR undoubtedly play an important role in this, international cooperation can, at least in theory, help to overcome some of these barriers (see Section 5.2). But if this is the case, why has progress been halting? And how does the multiplicity of institutions governing this field factor in? We return to these questions in Chapter 8, where we take stock of international institutions' effectiveness in governing FFSR to date.

5.7 References

Aldy, J. 2017. Policy Surveillance in the G-20 Fossil Subsidy Agreement: Lessons for Climate Policy. *Climatic Change* 144(1), 97–110.

APEC (Asia-Pacific Economic Cooperation). 2009. 2009 Leaders' Declaration. Singapore. Available at www.apec.org/Meeting-Papers/Leaders-Declarations/2009/2009_aelm.

APEC (Asia-Pacific Economic Cooperation). 2010. 2010 Leaders' Declaration. Available at www.apec.org/Meeting-Papers/Leaders-Declarations/2010/2010_aelm.aspx.

APEC (Asia-Pacific Economic Cooperation). 2011. 2011 Leaders' Declaration. Available at www.apec.org/Meeting-Papers/Leaders-Declarations/2011/2011_aelm.

APEC (Asia-Pacific Economic Cooperation). 2012. 2012 Leaders' Declaration. Available at www.apec.org/Meeting-Papers/Leaders-Declarations/2012/2012_aelm.

APEC (Asia-Pacific Economic Cooperation). 2013. 2013 Leaders' Declaration. Available at www.apec.org/Meeting-Papers/Leaders-Declarations/2013/2013_aelm.

APEC (Asia-Pacific Economic Cooperation). 2017. 2017 Leaders' Declaration. Available at www.apec.org/Meeting-Papers/Leaders-Declarations/2017/2017_aelm.

APEC (Asia-Pacific Economic Cooperation). 2018. About APEC. Available at www.apec.org/About-Us/About-APEC.

APEC (Asia-Pacific Economic Cooperation) Energy Ministers. 2010. 2010 APEC Energy Ministerial Meeting. Fukui, Japan. Available at https://apec.org/Meeting-Papers/Sectoral-Ministerial-Meetings/Energy/2010_energy.

APEC EWG (Asia-Pacific Economic Cooperation Energy Working Group). 2013. Guidelines on a Voluntary Peer Review for Reform of Inefficient Fossil Fuel Subsidies that Encourage Wasteful Consumption. Available at www.ewg.apec.org/documents/FINAL_VPR-IFFSR_Guidelines.pdf.

APEC EWG (Asia-Pacific Economic Cooperation Energy Working Group). 2015. APEC Fossil Fuel Subsidy Reform Capacity-Building Workshop: Summary Report. Available at www.ewg.apec.org/documents/FFSR_WS_Summary_Report_2-3-15.pdf.

APEC EWG (Asia-Pacific Economic Cooperation Energy Working Group). 2017. APEC Fossil Fuel Subsidy Reform Capacity-Building Workshop – Effective Pricing Mechanisms and Mitigation Strategies: Summary Report. Available at www.apec.org/-/media/APEC/Publications/2017/10/APEC-Fossil-Fuel-Subsidy-Reform–Capacity Building-Workshop-Effective-Pricing-Mechanisms-and–Mitigat/217_EWG_FFSR-WS2-Summary-Report.pdf.

Arze del Granado, F. J. and Coady, D. 2012. The Unequal Benefits of Fuel Subsidies: A Review of Evidence for Developing Countries. *World Development* 40(11), 2234–2248.

Asmelash, H. B. 2016. Falling Oil Prices and Sustainable Energy Transition. Towards a Multilateral Agreement on Fossil-Fuel Subsidies. United Nations University World Institute for Development Economics Research (WIDER) Working Paper. Available at www.wider.unu.edu/sites/default/files/wp2016-13.pdf.

Bast, E., Doukas, A., Pickard, S., van der Burg, L., and Whitley, S. 2015. *Empty Promises: G20 Subsidies to Oil, Gas, and Coal Production*. London/Washington, DC: Overseas Development Institute and Oil Change International.

Benes, K., Cheon, A., Urpelainen, J., and Yang, J. 2015. *Low Oil Prices: An Opportunity for Fuel Subsidy Reform*. New York: Columbia University Center on Global Energy Policy.

Climate Transparency. 2015. G20 Emissions. Available at www.climate-transparency.org/g20-emissions.

Coady, D., Parry, I., Sears, L., and Shang, B. 2017. How Large Are Global Fossil Fuel Subsidies? *World Development* 91(C), 11–27.

Eckersley, R. 2012. Moving Forward in the Climate Negotiations: Multilateralism or Minilateralism? *Global Environmental Politics* 12(2), 24–42.

Elek, A. 2005. Back to Canberra: Founding APEC. In *The Evolution of PECC: The First 25 Years*. Singapore: Pacific Economic Cooperation Council. Available at www.pecc.org/resources/regional-cooperation/601-back-to-canberra-founding-apec/file.

El-Katiri, L. and Fattouh, B. 2015. *A Brief Political Economy of Energy Subsidies in the Middle East and North Africa*. Oxford: Oxford Institute for Energy Studies.

Friends (Friends of Fossil Fuel Subsidy Reform). n.d. Events. Available at fffsr.org/events/page/3/.

Friends (Friends of Fossil Fuel Subsidy Reform). 2015. Fossil-Fuel Subsidy Reform Communique. Available at fffsr.org/wp-content/uploads/2015/06/friends-fossil-fuel-subsidy-communique_final.pdf.

Friends (Friends of Fossil Fuel Subsidy Reform). 2017. Implementing Paris and the SDGs through Fuel Subsidy Reform and Taxation – Country Best Practice. UNFCCC COP 23 side event description. Available at fffsr.org/event/implementing-paris-and-the-sdgs-through-fuel-subsidy-reform-and-taxation-country-best-practice/.

Friends (Friends of Fossil Fuel Subsidy Reform). 2018. Webinars. Available at fffsr.org/webinars/.

G20 (Group of 20). 2009. Leaders' Statement. Pittsburgh, United States. Available at www.g20.utoronto.ca/2009/2009communique0925.html.

G20 (Group of 20). 2012. Leaders' Declaration. Los Cabos, Mexico. Available at www.g20.utoronto.ca/2012/2012-0619-loscabos.html.

G20 (Group of 20). 2017. Hamburg Climate and Energy Action Plan for Growth. Available at www.g20.utoronto.ca/2017/2017-g20-climate-and-energy-en.pdf.

G20 Energy Sustainability Working Group. 2013. Methodology for G-20 Voluntary Peer Reviews on Inefficient Fossil Fuel Subsidies That Encourage Wasteful Consumption. Available at ru.g20russia.ru/load/783530379.

Garnaut, R. 2014. The G20 Has the Muscle to Fight Climate Change. The Australian Business Review. Available at www.theaustralian.com.au/business/business-spectator/the-g20-has-the-muscle-to-fight-climate-change/news-story/668174201c0a2d3d9c72807a94f88ec1.

Gehring, T. and Faude, B. 2014. A Theory of Emerging Order within Institutional Complexes: How Competition among Regulatory International Institutions Leads to Institutional Adaptation and Division of Labor. *The Review of International Organizations* 9(4), 471–498.

GSI (Global Subsidies Initiative). n.d. Friends of Fossil Fuel Subsidy Reform. Available at www.iisd.org/gsi/fossil-fuel-subsidies/friends-fossil-fuel-subsidy-reform.

IEA (International Energy Agency). 2017. *Tracking Fossil Fuel Subsidies in APEC Economies, toward a Sustained Subsidy Reform*. Paris: OECD and IEA. Available at www.iea.org/publications/insights/insightpublications/TrackingFossilFuelSubsidiesin APECEconomies.pdf.

IEA, OECD and World Bank. 2010a. Analysis of the Scope of Energy Subsidies and Suggestions for the G-20 Initiative. IEA, OPEC, OECD, World Bank Joint Report prepared for submission to the G-20 Summit Meeting Toronto (Canada), 26–27 June 2010. Available at www.oecd.org/env/45575666.pdf.

IEA, OECD and World Bank. 2010b. The Scope of Fossil-Fuel Subsidies in 2009 and a Roadmap for Phasing out Fossil-Fuel Subsidies: An IEA, OECD and World Bank. Report for the November 2010 G20 summit meeting in Seoul, Korea. Available at www.oecd.org/env/cc/46575783.pdf.

IEA, OECD and World Bank. 2011. Joint report by IEA, OPEC, OECD and World Bank on fossil-fuel and other energy subsidies: An update of the G20 Pittsburgh and Toronto Commitments. Prepared for the G20 Meeting of Finance Ministers and Central Bank Governors (Paris, October 14–15, 2011) and the G20 Summit (Cannes, November 3–4, 2011). Available at www.oecd.org/site/tadffss/49006998.pdf.

IISD (International Institute for Sustainable Development). 2016. Building on Momentum, Recommendations from the GSI for Fossil Fuel Subsidy Reform at the G20. Geneva: IISD. Available at www.iisd.org/sites/default/files/publications/building-on-momentum-recommendations-ffsr-g20.pdf.

IISD (International Institute for Sustainable Development). 2017. Members Call for WTO to Play a Central Role in Ending Inefficient Fossil Fuel Subsidies. Available at www.iisd.org/media/members-call-wto-play-central-role-ending-inefficient-fossil-fuel-subsidies.

Inchauste, G. and Victor. D. G. 2017. *The Political Economy of Energy Subsidy Reform*. Washington, DC: World Bank.

Jakob, M., Chen, C., Fuss, S., Marxen, A., and Edenhofer, O. 2015. Development Incentives for Fossil Fuel Subsidy Reform. *Nature Climate Change* 5(8), 709–712.

Jewell, J., McCollum, D., Emmerling, J., Bertram, C., Gernaat, D. E. H. J., Krey, V., Paroussos, L., Berger, L., Fragkiadakis, K., Keppo, I., Saadi, N., Tavoni, M., van Vuuren, D., Vinichenko, V., and Riahi, K. 2018. Limited Emission Reductions from Fuel Subsidy Removal Except in Energy-Exporting Regions. *Nature* 554(7691) 229–233.

de Jong, S. and Wouters, J. 2014. Institutional Actors in International Energy Law. In *Research Handbook on International Energy Law*, edited by K. Talus. Cheltenham: Edward Elgar.

Kim, J. A. and Chung, S. Y. 2012. The Role of the G20 in Governing the Climate Change Regime. *International Environmental Agreements: Politics, Law and Economics* 12(4), 361–374.

Kirton, J. J. and Kokotsis, M. E. 2015. *The Global Governance of Climate Change: G7, G20, and UN Leadership*. Farnham: Ashgate.

Koplow, D. 2018. Defining and Measuring Fossil Fuel Subsidies. In *Fossil Fuel Subsidies and Their Reform: The Politics of a Global Problem*, edited by J. Skovgaard and H. van Asselt. Cambridge: Cambridge University Press, 23–46.

Lemphers, N., Bernstein, S., and Hoffman, M. 2018. The Global Subsidies Initiative Catalytic Actors and the Politics of Fossil Fuel Subsidy Reform. In *The Politics of Fossil Fuel Subsidies and Their Reform*, edited by J. Skovgaard and H. van Asselt. Cambridge: Cambridge University Press.

Merrill, L., and Chung, V. 2015. *Financing the Sustainable Development Goals through Fossil-Fuel Subsidy Reform: Opportunities in Southeast Asia, India and China*. Geneva: GSI.

Merrill, L., Toft Christensen, L., and Sanchez, L. 2016. *Learning from Leaders: Nordic and International Best Practice with Fossil Fuel Subsidy Reform*. Copenhagen: Nordic Council of Ministers.

New Zealand Ministry of Foreign Affairs and Trade. 2010. Friends of Fossil Fuels Subsidies Reform Initiative. TND/NZE/24/3.

ODI (Overseas Development Institute) 2017. Germany Ignoring Majority of Fossil Fuel Subsidies in G20 Review – ODI Experts. Available at www.odi.org/news/840-germany-ignoring-majority-fossil-fuel-subsidies-g20-review-odi-experts.

OECD (Organisation for Economic Co-operation and Development). 2018. *OECD Companion to the Inventory of Support Measures for Fossil Fuels*. Paris: OECD.

Rentschler, J. and Bazilian, M. 2017. Reforming Fossil Fuel Subsidies: Drivers, Barriers and the State of Progress. *Climate Policy* 17(7), 891–914.

Reuters. 2017. G20 Urged to Ditch Fossil Fuel Subsidies by 2020, Go Green. Available at www.reuters.com/article/us-climatechange-subsidies-g-idUSKBN15U00Y.

Rive, V. 2016. Fossil Fuel Subsidy Reform: A New Zealand Perspective on the International Law Framework. *New Zealand Universities Law Review* 27(1), 73–101.

Rive, V. 2018. Anatomy of an International Norm Entrepreneur: The Friends of Fossil Fuel Subsidy Reform. In *The Politics of Fossil Fuel Subsidies and Their Reform*, edited by J. Skovgaard and H. van Asselt. Cambridge: Cambridge University Press.

Sanchez, L. 2017. Change Makers Leap Forward as Momentum for Fossil Fuel Subsidy Reform Grows. Available at fffsr.org/2017/05/change-makers-leap-forward-as-momentum-for-fossil-fuel-subsidy-reform-grows/.

Skovgaard, J. and van Asselt, H. 2018. The Politics of Fossil Fuel Subsidies and Their Reform: An Introduction. In *The Politics of Fossil Fuel Subsidies and Their Reform*, edited by J. Skovgaard and H. van Asselt. Cambridge: Cambridge University Press.

Smith, J. and Urpelainen, J. 2017. Removing Fuel Subsidies: How Can International Organizations Support National Policy Reforms? *International Environmental Agreements: Politics, Law and Economics* 17(3), 327–340.

Stefanski, R. 2014. *Dirty Little Secrets: Inferring Fossil-Fuel Subsidies from Patterns in Emission Intensities*. Oxford: University of Oxford.

Terton, A., Gass, P., Merrill, L., Wagner, A., and Meyer, E. 2015. *Fiscal Instruments in INDCs: How Countries are Looking to Fiscal Policies to Support INDC Implementation*. Geneva: International Institute for Sustainable Development.

UN (United Nations) 2015. Transforming our World: The 2030 Agenda for Sustainable Development. UN Doc. A/RES/70/1.

UNEP (United Nations Environment Programme) 2018. *Emissions Gap Report 2018*. Nairobi: UNEP.

UNEP (United Nations Environment Programme), OECD (Organisation for Economic Co-operation and Development), and IISD (International Institute for Sustainable Development). 2019. *Measuring Fossil Fuel Subsidies in the Context of the Sustainable Development Goals*. Nairobi: UNEP.

Van de Graaf, T. and Blondeel, M. 2018. Fossil Fuel Subsidy Reform: An International Norm Perspective. In *The Politics of Fossil Fuel Subsidies and Their Reform*, edited by J. Skovgaard and H. van Asselt. Cambridge: Cambridge University Press.

Van de Graaf, T. and van Asselt, H. 2017. Energy Subsidies at the Intersection of Climate, Energy and Trade Governance. *International Environmental Agreements: Politics, Law and Economics* 17(3), 313–326.

Van de Graaf, T. and Westphal, K. 2011. The G8 and G20 as Global Steering Committees for Energy: Opportunities and Constraints. *Global Policy* 2(1), 19–30.

Verkuijl, C., van Asselt, H., Moerenhout, T., Casier, L., and Wooders, P. 2019. Tackling Fossil Fuel Subsidies through International Trade Agreements: Taking Stock, Looking Forward. *Virginia Journal of International Law* 58(2), 309–368.

Victor, D. G. 2009. *The Politics of Fossil-Fuel Subsidies*. Geneva: Global Subsidies Initiative.

Wade, R. H. 2011. Emerging World Order? From Multipolarity to Multilateralism in the G20, the World Bank, and the IMF. *Politics & Society* 39(3), 347–378.

World Bank. 2014. Transitional Policies to Assist the Poor While Phasing Out Inefficient Fossil Fuel Subsidies That Encourage Wasteful Consumption. Contribution by the World Bank to G20 Finance Ministers and Central Bank Governors. Available at www.g20.org/sites/default/files/g20_resources/library/14%20Transitional%20Policies%20To%20Assist%20The%20Poor%20While%20Phasing%20Out%20Inefficient%20Fossil%20Fuel%20Subsidies.pdf.

Young, M. A. 2017. Energy Transitions and Trade Law: Lessons from the Reform of Fisheries Subsidies. *International Environmental Agreements: Politics, Law and Economics* 17(3), 371–390.

6

Carbon Pricing

Overlaps and Formal Collaboration

JAKOB SKOVGAARD AND JANA CANAVAN

6.1 Introduction

Putting a price on carbon provides a straightforward instrument for climate policy, but it also has important repercussions for energy use. This is because most emissions covered by carbon pricing and markets stem from industries with high energy use, and because carbon prices suppress the consumption of energy through directing the choice of fuels away from emissions-intensive fuels. A global price on carbon has been touted as *the* solution to climate change by actors across the political and geographical spectrum, especially economists (Ball 2018).

It is no surprise then that the last ten years have seen a surge in international and transnational institutions aimed at promoting carbon pricing and carbon markets. A couple of such institutions have existed since the 1990s (most notably the International Emissions Trading Association, IETA), but most have appeared since 2007 (Sanderink et al. 2016). These institutions have promoted carbon taxes and emissions trading, as well as systems for the offsetting of emissions. More specifically, the general promotion of placing a price on carbon has taken its shape in the form of the setting of standards and commitments, information-sharing and networking, operational activities such as pilot and demonstration projects, and, to a lesser degree, financing.

The overall purpose of this chapter is to provide an overview of the existing carbon pricing and trading institutions. Specific attention lies on mapping out their focus areas and points of interaction that shape the roles, areas of specialization, and underlying norms that relate to the pricing of carbon emissions. We illustrate that carbon-pricing institutions constitute a subfield of interconnected and interactive parts, which together perform crucial tasks of carbon taxing, emissions trading, and offsetting; all directed toward promoting wider carbon-pricing efforts. In this, we follow the argument of Sanderink et al. (2016), as well as Zelli et al., and Sanderink et al. in Chapters 2 and 3, that it is instructive to identify the membership, governance functions, and interlinkages between institutions within such a subfield of the climate-energy nexus.

The institutions we target for such an analysis constitute global (public and hybrid) institutions that focus on promoting carbon pricing on the international level and are vital to the wider field of climate change governance. While no orchestrating entity exists, we hold that the United Nations Framework Convention on Climate Change (UNFCCC) and carbon-pricing institutions embedded in the World Bank (the Carbon Pricing Leadership Coalition [CPLC], the Networked Carbon Markets [NCM], and the Partnership for Market Readiness [PMR]) constitute the most important institutions within the subfield. The UNFCCC and the World Bank differ in that the UNFCCC is an environmental institution that constitutes the incumbent and the central hub within the climate nexus, while the World Bank's involvement with climate change is more peripheral but has been increasing during the last twenty years (Park 2010; Gallagher and Yuan 2017). Beyond the UNFCCC and the World Bank institutions, a range of public, hybrid, and especially private institutions also constitute parts of the carbon-pricing subfield. Public institutions included here are the Western Climate Initiative (WCI), the International Carbon Action Partnership (ICAP), the Carbon Offsetting and Reduction Scheme for International Aviation (CORSIA), and, as mentioned, the UNFCCC and the PMR. Hybrid institutions cover the UN Global Compact for Climate (C4C), the NCM, and the CPLC. Lastly, covered private institutions are the Gold Standard, Carbon Neutral Protocol (CNP), Verified Carbon Standard (VCS), the (IETA), and the International Air Transport Association Carbon Offset Program (IATA_COP).

So far, studies of carbon pricing have mainly focused on economic aspects and on single cases of carbon-pricing efforts at national, provincial, and European Union levels (Skjærseth and Wettestad 2008; Harrison 2012; Sterner and Coria 2012; but see Betsill and Hoffmann 2011). Several scholars have analyzed carbon markets from a critical perspective and emphasized their neoliberal underpinnings (Stephan and Paterson 2012; Lane and Newell 2016). Analyses covering both carbon taxes and emissions trading are rare, as are comparative studies of the adoption of carbon pricing (Harrison 2012; Rabe and Borick 2012). Concerning the international level, studies of the diffusion of carbon-pricing instruments, such as carbon markets, tend to focus on the diffusion between peers (from government and industry) in different polities, while paying less attention to the role of international institutions that promote such diffusion (Meckling 2011a; 2011b; Stephan and Paterson 2012; Paterson et al. 2014).

This chapter will contribute to this literature by focusing on the neglected issue of what the governance of carbon pricing on the international level looks like. In the same vein, the chapter contributes to the analytical ambition of this book by exploring and mapping out the meso level of international institutions that are promoting carbon pricing.

The existence of no less than thirteen institutions promoting carbon pricing begs the question of how the institutions align in terms of membership, governance functions, and their interpretation of the norm of carbon pricing, as well as how they interact, especially given that there is no immediately visible division of labour or orchestrating entity. The alignment and interaction are particularly important given the interdependence among these institutions in their efforts to promote carbon pricing. This interdependence is also rooted in the core norm of carbon pricing, which is based on the notion that climate change should be addressed through placing a price on emissions corresponding to its social costs (Nordhaus 2008). All institutions within the carbon-pricing subfield subscribe to this core norm of the subfield, yet there seems to be differences in terms of how this idea has been interpreted in practice (Meckling and Jenner 2016), which may have implications for the legitimacy and effectiveness of the work that they undertake.

The fact that the core norm has been interpreted in diverging ways and that the subfield is characterized by a medium number of institutions (fewer than the institutions addressing renewable energy, more than those addressing fossil fuel subsidies) also implies that it is difficult to predict the degrees of coherence and management of the subfield. Scholars of institutional fragmentation and polycentricity have argued that a shared core norm can enhance the coherence across a governance system (Ostrom 1990; Biermann et al. 2009). Yet, the diverging interpretations of the core norm may qualify such an effect. The same goes for potential cross-institutional variations of other dimensions such as membership or governance function. All this suggests to study the actual degree of coherence and management of the carbon-pricing subfield in greater detail.

Furthermore, it is worth mapping the subfield of carbon-pricing institutions due to their place within the climate-energy nexus complex (see Chapter 1). Carbon pricing, unlike fossil fuel subsidy reform and renewable energy, is explicitly climate-focused. Consequently, the UNFCCC plays a central role among the institutions that promote carbon pricing, but there is also a plethora of other institutions with little or no relation to the UNFCCC that have been highly active in the promotion of carbon pricing. We therefore explore the overall level of coherence between the institutions, and zoom in on the dyadic interlinkage between the UNFCCC and institutions embedded in the World Bank.

Following the analytical framework laid out in Chapter 2, we examine to which degree the interlinkage of said institutions is characterized by coherence, and how the degree of coherence has been managed by the institutions. Methodologically, this chapter is based on a qualitative case study that involves documentary analysis of official documents and secondary sources, as well as targeted interviews with key officials previously or currently working for the secretariats of the institutions under scrutiny or for closely related international organizations. All in all, twenty semi-structured interviews were conducted, either in person or via phone and Skype, whilst being audio-recorded, transcribed, coded, and analyzed with the NVivo programme. The interview questions as well as the coding of the interviews focused on the role of the individual institutions, their interlinkages and the attempts to manage them, and the overall subfield, as well as how the norm of carbon pricing has been interpreted.

Our chapter will proceed with first providing an overview of carbon pricing as a policy instrument, including a discussion of carbon taxes, carbon markets, and offsets as well as a review of the literature on carbon pricing. Subsequently, the international institutions that promote carbon pricing and carbon markets are mapped in terms of their interpretation of the previously mentioned core norm, their membership (public, private, or hybrid), and their governance functions. After this meso-level analysis, the chapter zooms into the interlinkage between the UNFCCC and the World Bank institutions and on how the interlinkage between these institutions has been managed.

6.2 Carbon Pricing: An Economic Solution to an Environmental Problem

Pricing greenhouse gas emissions is the fundamental solution to climate change, according to much of the environmental economics literature (see, for instance, Jacobs 1997; Tol 2011; Sterner and Coria 2012). Institutions that promote such pricing of emissions can be said to reflect the norm complex of liberal environmentalism (Bernstein 2001), and carbon markets in particular have been described as a key component of 'climate capitalism' (Lane and Newell 2016). Consequently, these institutions have been promoted by leading economists, economic organizations such as the International Monetary Fund and the World Bank, and influential journals and newspapers such as the Economist or Financial Times, as well as environmental NGOs such as the World Wildlife Fund (WWF).

Carbon pricing can take form through carbon taxation or the trading of allowances to emit greenhouse gases in a carbon market. The term 'carbon market' refers to systems for trading with other entities that are covered by the same emissions trading or cap-and-trade system with an overarching cap. Moreover, the term covers systems for purchasing carbon credits (or 'offsets') from entities outside of said target and that can be counted toward an emissions target, e.g. the Clean Development Mechanism (CDM) (Paterson et al. 2014).

Measures of carbon pricing were first adopted by Nordic countries such as Finland (1990) and Sweden (1991), which introduced carbon taxes preceding international agreements on climate change. Throughout the 1990s, carbon taxes were adopted by (mainly smaller) European countries. Carbon markets, which were inspired by US experiences with creating a market for trading allowances to emit sulphur (Aldy and Stavins 2012), quickly became one of the most popular climate-policy instruments in the period following the adoption of the Kyoto Protocol (Meckling 2011a; Meckling 2011b; Paterson 2012). In terms of covered emissions, the EU emissions-trading system introduced in 2005 constitutes the largest carbon-pricing instrument in the world and was crucial in establishing carbon markets as a key climate policy instrument. Private corporations (e.g. British Petroleum) as well as local and regional governments (e.g. California) also adopted carbon-pricing instruments.

More recently, however, carbon taxes have regained some of the attention they had received in the early and mid-1990s, while carbon markets (especially the EU emissions-trading system and the offset markets) have been plagued by periods of falling demand and prices. Both instruments have since 2010 been adopted by a diverse set of countries covering all regions of the world and different political systems and levels of income (Skovgaard et al. 2019). The emissions covered by carbon-pricing policies across the globe encompass 20 per cent of global emissions, mainly stemming from energy use within industry, transportation, and power generation, whereas emissions from non-energy use (e.g. agriculture, forestry, or waste) are covered in very few cases (World Bank 2018b).

6.3 Meso-Level Coherence

In this third section of the chapter, we will map out the field of international carbon-pricing institutions that are anchored around the World Bank and the UNFCCC. The first subsection is dedicated to provide a short overview of the emergence of the institutional complex on carbon pricing, followed by a fourfold distinction of how the institutions under scrutiny interpret the core norm underlying this subfield. Next, we describe the patterns of memberships and governance functions that shape the resulting net and coalitions among the selected institutions.

6.3.1 Emergence of the Institutional Complex on Carbon Pricing

The first cases of carbon pricing occurred long before the establishment of international institutions that would support such efforts. It was only following the introduction of carbon markets in the Kyoto Protocol that the first institutions were introduced specifically to promote carbon pricing. These were transnational business coalitions, most noteworthy IETA, that cover particularly finance and energy corporations and environmental NGOs, and are often highlighted as a key factor in the early diffusion of carbon markets (Meckling 2011a; Paterson 2012). These early carbon-pricing institutions promoted carbon markets as *the solution* to climate change and were important in the adoption of carbon markets in the European Union and US states; particularly in California and the north-eastern states (Meckling 2011a; Paterson 2012). The UNFCCC played an important role both in defining carbon markets as a key policy instrument in the Kyoto Protocol and in subsequently promoting and defining the rules for offsets within the Kyoto Protocol framework (see discussion of the offsets in Section 6.4.1).

Yet, most of the carbon-pricing institutions that are currently active have been established since 2007. Recently, institutions promoting carbon pricing have proliferated, rather than solely carbon markets. This indicates a significant, yet under-explored, new development in the international governance of carbon pricing. It is important to note that this surge in carbon-pricing institutions involves public actors to a much larger degree than the initial carbon-market promoters, as discussed in the sections to follow. As we will show, the differences among these approaches reflect variations in how the underlying core norm of reducing emissions through pricing is interpreted.

6.3.2 The Core Norm of Carbon Pricing

The core norm of carbon pricing is based on the notion that climate change is best mitigated by giving emitters an incentive to reduce emissions in terms of a price signal, and that the decision of how to reduce emissions is best left to the market. These notions are, in turn, underpinned by the understanding of actors as economically rational, and of the response to climate change as compatible with liberal and capitalist systems.

Yet, while this core norm is fundamental to all carbon-pricing policies and efforts to promote carbon pricing, it can in practice be interpreted in rather diverging ways. In an institutional complex in which several institutions with different memberships are embedded within the climate-energy nexus, there is scope for diverging and even conflicting applications of this norm. We therefore distinguish four dimensions along which interpretations of the core norm may vary: quantity versus price instruments, whether polluters should pay for all of their emissions or not, mandatory versus voluntary schemes, and carbon pricing within a given jurisdiction versus offsetting. For each dimension, we discuss how the institutions have interpreted the dimension in practice.

First, on the most basic conceptual level, there is a key distinction between placing the costs of the externality of climate change on the polluters (Pigou 1932; Jacobs 1997) – thus also adhering to the ‘polluter pays principle’

(OECD 1974) – and between creating a system to allocate property rights to emit greenhouse gases, as well as to the trading of these rights (Coase 1960; Felli 2015). While carbon taxes explicitly constitute taxation, the tax component of carbon markets with auctioning consists of the money that polluters have to pay to the state (or other auctioning entity) for each emission allowance, and is hence more implicit. Nonetheless, this tax component is easily identifiable to the industry sectors that have to purchase the allowances (Rabe and Borick 2012).

Furthermore, carbon markets regulate the quantity of emissions, while carbon taxes regulate the price of emissions. This distinction between quantity and price instruments led Meckling and Jenner (2016) to argue that whereas carbon markets are rooted in neoliberalism and a US-dominated tradition of policy making, carbon taxes are rooted in ordo-liberalism and the European policy-making tradition. According to Meckling and Jenner, the former tradition delegates more authority to market forces since it creates a new regulatory market. Yet, we are not convinced by the association between carbon taxes and ordo-liberalism, since Pigou was a Keynesian economist and since carbon taxes are preferred over carbon markets by all kinds of economists worldwide (including neo-classical ones). This is due to its more direct imposition of the externality on the polluter (Rabe and Borick 2012).

While all institutions have promoted carbon markets, carbon taxes have almost solely been established by public and especially hybrid institutions, most notably the CPLC and the PMR. This is unsurprising given that the private institutions in question have been established to promote functioning carbon markets (with the exception of the C4C, which advocates that companies set an internal shadow carbon price). A more interesting development is an apparent move away from focusing almost solely on carbon markets to increasingly promoting carbon taxes in parallel. We will discuss this development further when focusing on the World Bank institutions in a subsequent section.

Second, there is a distinction between whether polluters must pay for all of their emissions – as they do in systems with a carbon tax and in emissions-trading systems in which all allowances are auctioned – or whether polluters only pay for emissions above a given baseline – as they do in emissions-trading systems with free allocation (so called grandfathering; see Aldy and Stavins 2012) and in case of voluntary offsets. These two options constitute parts of a continuum, with several carbon-market policies operating somewhere in between. For example, most of the world's emissions-trading systems combine grandfathering and auctioning of allowances. Mandatory carbon taxes nonetheless always imply that all emissions are subject to the polluter paying for them. In terms of concrete interpretation, most of the institutions do not hold an explicit official position in this regard. This may be explained by the fact that a bulk of institutions that promote carbon markets would meet less support for their efforts if they explicitly preferred full pricing of all emissions.[1]

The third conceptual dimension is whether carbon pricing is mandatory or voluntary. Carbon pricing has been adopted either in the form of mandatory schemes that cover all entities within particular sectors operating within the polity (states, sub-national entities such as provinces, and supranational entities such as the EU), or as voluntary schemes (mainly carbon markets) joined by companies that would like to commit to reducing or offsetting their emissions. Unlike most other mitigation policies, mandatory carbon pricing provides revenue for the public budget, a characteristic appealing to powerful finance ministries and politicians facing budgetary constraints. Voluntary carbon markets, on the other hand, refer to institutionalized markets that are responsible for trading those verified emissions reductions (VERs) that are not part of the regulatory schemes under the Kyoto Protocol and the EU ETS (Benwell 2009; Segerson 2013).

The voluntary carbon-trading actions of this sector are thus constituted of the activities of organizations or individuals taken outside of, in addition to, or beyond the existing environmental policies or basic environmental laws and regulations on carbon emission and trading. Operating independently from the UNFCCC emission targets and offset mechanisms, the voluntary carbon-trading markets are led by various public and private actors and follow standards created by its industrial stakeholders. Besides offering opportunities to engage in emissions trading and to enable genuine reduction of carbon emissions that could potentially exceed the goals set by mandatory carbon-trading markets, there are several other motivational factors for engaging in voluntary carbon-trading measures. In reaction to the normative pressure from NGOs or externally existing regulations to reduce emissions, actors can use carbon-trading measures to fulfil corporate social responsibility (CSR) goals and to realize marketing opportunities in line with liberal environmentalist goals and values (Lyon and Maxwell 2007; Benwell 2009). The voluntary carbon market therefore plays an influential role for the private sector as it focuses on individual consumers and green consumerism (Lyon and Maxwell 2007; Choi 2015). However, voluntary schemes are often established in relation to existing regulatory schemes, which means that they may undermine the process of establishing successful mandatory policies (Lyon and Maxwell 2003; Segerson 2013).

The institutions analyzed here have not explicitly taken a stance on whether mandatory or voluntary approaches are preferable. Most public and hybrid institutions (except the C4C) work mainly with mandatory policies, whereas

[1] Interview with senior official from the NCM, 25 May 2017; interview with senior official from the CNP, 28 July 2017.

IATA_COP and, to some degree, the Gold Standard and the NCM work with voluntary carbon markets or voluntary offsets. The different involvement in either regulatory or voluntary carbon pricing reinforces the divide between public actors such as the UN, the World Bank Group, and closely related state regulations and policies on the one hand, and the private actors referring to business corporations and individual consumers on the other hand. Within some areas, most notably aviation, there has been a development to move from voluntary standards (IATA_COP) to mandatory ones (CORSIA).

The fourth dimension of the norm of carbon pricing relates to the distinction between carbon taxes and emissions trading that reduce emissions within a defined jurisdiction on the one hand, and offsetting on the other hand, which enables the purchasing of carbon credits (or 'offsets') from entities in other jurisdictions, for example as in the CDM. Whereas both taxational emissions trading and monetary offsets are referred to as constituting carbon markets due to their shared focus on operating through the trading of emissions allowances, they differ regarding this key distinction of jurisdictions.

Thus, none of the institutions have explicitly endorsed offsets over within-jurisdiction reductions, or vice versa. In their practices, however, they have generally promoted one or the other. Today, offsets are to a larger degree supported and disseminated by private institutions, rather than by public and hybrid ones. Nevertheless, none of the latter are, as such, opposed to them. Offsets such as the CDM were defined as a key instrument in the global response to climate change by the UNFCCC and the Word Bank institutions in the years between the Kyoto Protocol and the Copenhagen Accord. Importantly however, the focus has increasingly turned to the linking of carbon markets, especially in the context of Article 6 of the Paris Agreement (Kansy 2016).[2]

In summary, while all institutions promote the norm of carbon pricing, there are important differences in how they interpret the norm in practice, creating clusters of private institutions on the one hand, and public and hybrid institutions on the other. These differences were most pronounced regarding the choice between carbon markets and carbon taxes.

6.3.3 Membership

In this section, we will outline the membership of the carbon-pricing and carbon-market institutions. This allows us to map how the institutions differ in terms of coverage of actors that have diverging preferences from the members of another

[2] The Paris Agreement does contain a provision establishing a new 'sustainable development mechanism' (Article 6.4), which will constitute a new kind of offset mechanism oriented not only toward trading emissions allowances but also promoting sustainable development beyond climate change.

Table 6.1 *Overview of governance functions across different types of institutions (public, hybrid, private) for carbon pricing.*

	Public	Hybrid	Private
Standards & Commitments		UN Global Compact Caring for Climate (C4C)	Gold Standard Carbon Neutral Protocol (CNP) Verified Carbon Standard (VCS)
Operational Activities	Western Climate Initiative (WCI)		
Information & Networking	International Carbon Action Partnership (ICAP)	Networked Carbon Markets Initiative (NCM) Carbon Pricing Leadership Coalition (CPLC)	International Emissions Trading Association (IETA)
Standards & Commitments; Operational	Carbon Offsetting and Reduction Scheme for International Aviation (CORSIA)		International Air Transport Association Carbon Offset Program (IATA_COP)
Information & Networking; Financing	Partnership for Market Readiness (PMR)		
Standards & Commitments; Information & Networking	United Nations Framework Convention on Climate Change (UNFCCC)		

institution, e.g. business and state actors. The carbon-pricing issue area mapped out here consists of thirteen institutions (see Table 6.1), comprising public, private, and hybrid constituencies. While a few of them have existed since the 1990s, most have been established from 2007 onwards. Business and public actors (states, IOs, and sub-national governments) are the main constituents, while civil society organizations are only involved in the CPLC, the C4C, the Gold Standard, and the NCM. With the exception of the WCI (which covers the states and provinces on the West coast of the USA and Canada), all institutions are global in terms of membership and reach. However, their members (especially from business and CSOs) tend to be concentrated in industrialized countries and, to a lesser degree, emerging economies.

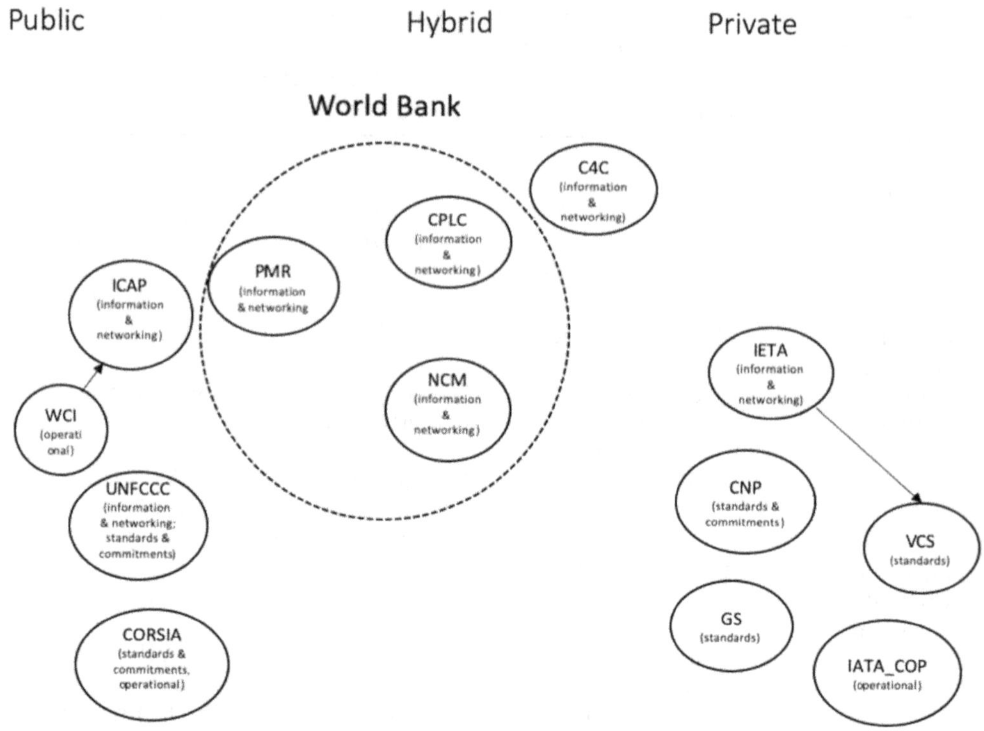

Figure 6.1 Public, hybrid, and private carbon pricing institutions and their membership relations.

One important feature of carbon market institutions is, thus, that several of them are related to, or nested within, wider institutions. For instance, the C4C is nested in the UN Global Compact, which is a UN initiative to motivate businesses to adopt sustainable and socially responsible policies, while the PMR, the NCM, and the CPLC are nested within the World Bank (see also Figure 6.1). Regarding the aviation sector, CORSIA, adopted in 2016, is nested within the public International Civil Aviation Organization (ICAO), whereas IATA_COP is nested within IATA, which is the international trade association for the aviation industry. Institutions that are not couched within wider overarching institutions were often established by other international institutions, e.g. IETA being one of the founders of VCS, and the Gold Standard being founded by the WWF. Thus, some of the relations between the institutions are not only very close, but also hierarchical in nature (Alter and Meunier 2009).

Figure 6.1 depicts some of these connections and also provides a first general overview of the institutions' governance functions, which will be discussed in further detail in Section 6.4. The figure pictures the membership relations whilst being organized under the categories of public, hybrid, and private. Arrows denote that the institution is a member of the institution that the arrow points to. Note that the institutions' position within the public, hybrid, and private groups, respectively, does not implicate a hierarchy or that they are in other ways more or less 'public' or private than other institutions within the same group.

6.3.4 Governance Functions

Mapping out the membership and governance functions of the thirteen institutions shows that they differ to some degree in their focus on various carbon-market activities, but that there are also considerable geographical and functional overlaps. In this section, we therefore identify and discuss two major clusters within the issue area; one centred around the public and hybrid institutions embedded in the World Bank, and another consisting of private institutions, mainly centred around IETA (see Figure 6.1). In order to illustrate our chosen approach to clustering the institutions, we later zoom in on two sets of public and hybrid institutions, namely the UNFCCC and the World Bank institutions, as we find them more politically and academically relevant than the private institutions (Section 6.4).

The **private institutions** have been instrumental in promoting emissions trading globally and in different polities around the world (Meckling 2011b; Paterson 2012; Paterson et al. 2014). Their key objective is to further carbon markets, which offers companies (including financial companies) useful business opportunities in the growing carbon-market sector. At the same time, they give emitting industries the possibility to continue their activities without costly carbon taxes and regulation (Paterson 2012). The Gold Standard's efforts to ensure and improve the environmental and social integrity of offsets stands out in this respect, due to its clear and comprehensive focus on supporting social aspects of sustainable development.

In terms of governance functions, the private institutions focus on providing a venue for information and networking (IETA), on setting standards and commitments for offsets (VCS, Gold Standard), and on advancing company carbon neutrality (CNP). In order to facilitate successful and validated greenhouse gas emissions trading, actors such as CNP and VCS offer businesses, organizations, and technical partners a global standard framework for achieving carbon neutrality through internal mitigation measures (e.g. energy efficiency) and emission offsetting.

The goal of most of these private institutions is to achieve carbon neutral economic growth. For example, IATA_COP, a leading carbon offset programme for the aviation industry, endorses voluntary offset schemes in which passengers pay to offset the emissions caused by their individual share of the flight's emissions (IATA 2009). Investing in such voluntary efforts to combat climate change with a focus on the individual customers allows businesses to address the CO_2-emissions impact of their industry without having to suppress the demand for air travel, and diminishes the call for mandatory and public regulation.

The governance functions of **public institutions** mainly regard the provision of information and networking possibilities between carbon-pricing actors, particularly from countries that have, or are considering to, implement carbon-pricing policies. The UNFCCC and CORSIA also engage in setting standards and commitments (both described subsequently). Of the public institutions, CORSIA and the WCI engage in operational activities, while the PMR provides financing for polities interested in adopting carbon-pricing policies. The aim of these undertakings is generally to offer platforms for collaboration to achieve decarbonization of specific sectors, countries, or states within federal systems. The goal of ICAP, for instance, is to provide a platform to strengthen the compatibility and effectiveness of the regulated carbon-trading market in order to promote innovation and allow for ambitious global reductions of global warming emissions. The objective of public institutions such as ICAP, the PMR, and the WCI is thus to provide a platform to strengthen the compatibility and effectiveness of carbon pricing. Following dissatisfaction with the private, voluntary efforts to curb aviation emissions, public and mandatory regulations to reach carbon neutrality were introduced with the establishment of CORSIA in 2016.

In public institutions, both carbon markets and taxes are promoted, albeit with a stronger emphasis on promoting carbon markets. Important in this respect is the objective of creating a global carbon market or at least to link the different carbon markets. Such linking is believed to improve economic efficiency by ensuring uniform prices and thus avoid distorting competitiveness and utilize low-cost abatement options (Kansy 2016). For instance, PMR-led programmes provide countries with grant funding to support the implementation of carbon taxes or emissions trading and include programmes for technical and policy work. The most recently established institution, CORSIA, has been created by the public aviation institution ICAO to ensure the offsetting of emissions above a given level; here relating to the total emissions of global aviation in 2020. CORSIA is voluntary until 2027, after which participation becomes mandatory for all countries except for those with a very low share of global aviation or those that are most vulnerable to climate change due to poverty and other structural inequalities.

Hybrid institutions perform governance actions and services that largely seek to bridge the gap between the public regulatory and the private voluntary markets to achieve broader and globally applicable schemes. The NCM, for instance, aims to support various actors from civil society, governments, and the private sector to link voluntary and mandatory carbon markets. Their goal is thus to facilitate cross-border trade and link carbon markets through improving the transparency and comparability of the existing markets (NCM 2017). The CPLC, too, offers voluntary partnership for leaders across governments, the private business sector, and civil society who share the long-term goal of achieving a global carbon-pricing economy.

The major governance functions of hybrid institutions include information-sharing, networking, capacity building, and knowledge-sharing, which are viewed as necessary instruments to connect strong institutions and regulations with the workings of the market economy. The CPLC and the C4C, for example, view carbon pricing as an essential step to approach zero net emissions, and both institutions see their core role in forming coalitions to approach a global carbon-pricing economy.

The institutions we studied are, with the exception of the WRI, all global in scope. Apart from CORSIA and IATA_COP, which focus on aviation, none of these institutions have specific sectoral foci, but focus on mitigation in general. De facto, this implies that the institutions address emissions from energy use to a significant degree, since virtually all carbon-pricing measures address energy use while not many of them cover emissions from other sources.

When summing up the just-presented mapping of institutional governance functions, it can be said that although all thirteen institutions work toward mitigating climate change by placing a price on carbon emissions, they do differ in regard to which activities they support. Broadly speaking, the public and hybrid institutions mainly focus on the support of *political decisions* to implement carbon pricing (e.g. CPLC, PMR) and to link mandatory carbon markets (e.g. ICAP, NCM), whereas the private institutions tend to focus on the facilitation of the *trading* of emissions allowances between private entities (e.g. IETA, see subsequent discussion). Among the private institutions we examined, both the Gold Standard and VCS aim to improve the social and environmental integrity of offsets. IETA[3] is an association for companies within the carbon-market sector that works for a functional international framework for carbon trading, whereas IATA_COP offers voluntary offsets for air travel.

6.3.5 Summary: Coherence at the Meso Level

When assessing the overall consistency of the thirteen carbon-pricing institutions, one needs to consider that there is no clear division of labour, as their functions and

[3] IETA also comprises the International Carbon Reduction and Offset Alliance (ICROA), which sets standards for voluntary offsets.

activities overlap to a large degree. Not only do most of them cover the same global geographical scope and the same policy sectors (although CORSIA and IATA_COP focus on international aviation), but they often also perform similar governance functions with similar objectives. It can thus generally be said that the institutional carbon-pricing complex is characterized by a medium level of consistency with some duplication and coexistence.

Regarding the application of the core norm, one can observe medium consistency, with several of the private institutions promoting voluntary carbon markets and offsets, while public institutions promote mandatory carbon pricing in the shape of taxes as well as carbon markets. Yet, these differences in application do not inherently involve conflict. In terms of membership, we also see a medium degree of consistency, with considerable overlap between the memberships of the different institutions – and without any organizing principles except for institutions being respectively public, hybrid, and private.

Finally, concerning governance functions, there are considerable overlaps in terms of undertaking similar functions (especially Information and Networking, see Table 6.1) on a global level, but also a divergence that ensured that most governance functions were covered. In many cases, public, private, and hybrid institutions are performing similar governance functions to achieve similar or related objectives. For instance, ICAP (public), the NCM (hybrid), and IETA (private) all work to promote the linking of carbon markets by providing information and networking opportunities. They thus overlap in terms of what they do (governance functions and carbon markets) but represent different members with potentially diverging preferences, e.g. IETA representing the interests of the carbon-market sector and ICAP those of the polities with emissions-trading policies.

This notwithstanding, our interviews show that collaboration is more common than competition. The widespread information and networking activities (e.g. conferences and workshops) provide ample opportunities for maintaining informal personal contacts.[4] Furthermore, sometimes institutional interlinkages are formalized in terms of Memoranda of Understanding or other written agreements between institutions, such as the Memorandum of Understanding between the IATA_COP and IETA. Importantly, formal relationships also exist in terms of institutions being members of each other (e.g. the WCI being a member of ICAP) or through participating in meetings of major institutions (e.g. of the UNFCCC). The World Bank institutions also often act as central hubs for interlinkages since they can draw on the expertise of the different World Bank departments.[5]

[4] Interview with VCS official, 31 May 2017.

[5] Interview with NCM official, 22 May 2017, Interview with NCM and PMR official, 25 May 2017.

While the large amount of institutions may indicate possible competition, especially when new initiatives enter the field, our analysis shows that the here-mapped institutions often seek to avoid this by informal communication and networking, e.g. on the individual level.[6] Through bilateral or multilateral exchange, policy makers engage in technical dialogue on the operation of their carbon markets and opportunities for deepening and connecting those markets.[7] Hence, situations that could potentially lead to conflict have been defused through behavioural mechanisms of ad-hoc coordination. Furthermore, the institutions often collaborate on operational activities, e.g. the implementation of offset projects, and institutionalized benchmarking, for instance in making sure that CORSIA will only use credits that followed the standards of VCS or other institutions.[8]

However, the most predominant mechanism for collaboration includes knowledge production and leadership to drive ambition. Key activities include organizing workshops for stakeholders and co-developing guidance materials to mainstream climate-leadership practices. All collaborations are multi-stakeholder in nature, involving companies large and small as well as civil society and public institutions.[9] Often such collaboration is based on complementarity, e.g. the diverse areas of expertise that the institutions exhibit.[10] It equally builds on their diverse membership circles, inasmuch as private institutions often collaborate with public and hybrid institutions.[11] Such divisions of labour are generally not based on official agreements as much as on informal assessments of relative strengths.[12]

Altogether, the field is characterized by some duplication but also bilateral (sometimes ad-hoc) coordination. Arguably, had the institutions differed more on the core norm, e.g. if not all of them were in favour of carbon markets, coordination could not have played the same role. The conducted interviews indicate that the level of coherence can be explained due to informal contacts and a desire to avoid overlaps, rather than the institutions being synergetic by design.

6.4 Micro-Level Coherence

6.4.1 Institutions under Scrutiny

We argue that the UNFCCC and the World Bank–embedded institutions constitute the most politically important institutions within the subfield of carbon pricing, and two incumbent nodal institutions of the climate-energy nexus as a whole. Apart

6 Interview with senior PMR official, 27 August 2018.
7 Interview with senior ICAP official, 23 May 2017.
8 Interview with VCS official, 31 May 2017.
9 Interview with NCM and PMR official, 25 May 2017.
10 Interview with Gold Standard official, 22 May 2017; Interview with NCM and PMR official, 25 May 2017.
11 Interview with C4C official, 20 June 2017; Interview with IATA official, 25 May 2017.
12 Interview with senior PMR official, 27 August 2018.

from their empirical importance, their crucial differences also give them high theoretical relevance. The UNFCCC is an intergovernmental environmental institution based on a multilateral environmental agreement, which serves as the forum for further intergovernmental negotiations concerning how to address climate change on the global level. The World Bank, on the other hand, is an economic institution, more specifically a multilateral development bank, focusing on poverty eradication and building shared prosperity in developing countries (Nielson and Tierney 2005; Park 2005). These thematic differences notwithstanding, both the UNFCCC and the World Bank, together with their secretariats and embedded institutions, share certain characteristics, such as their global membership reach and their public or hybrid (the CPLC and the NCM) constituencies.

Not unlike the way in which the UNFCCC Secretariat supports the UN climate regime, the World Bank provides secretariat services to the PMR, the NCM, and the CPLC. These services are managed by the World Bank's climate change group, sometimes with the same person working for more than one institution. We refer to this form of relationship as embeddedness within the World Bank, although the associated institutions differ in their nature, with the CPLC being a coalition, the PMR a trust fund, and the NCM a World Bank initiative. Physically located at the World Bank headquarters in Washington, DC, the three World Bank institutions differ in their roles. The CPLC is a coalition of actors from business, civil society, and politics with the purpose of advocating carbon pricing and, increasingly, promoting carbon pricing among businesses. The PMR is a World Bank Group multi-donor trust fund that provides technical advice and funding to the (at the time of writing) nineteen developing countries that are interested in developing carbon-pricing policies. None of these countries are low-income countries. The trust fund also seeks to create and share knowledge about carbon pricing.[13] The members of the PMR are state governments, usually represented by UNFCCC negotiators. The World Bank is also the trustee and the delivery partner of the PMR. Finally, the NCM supports the linking of climate markets through ensuring that the tradeable units from the different markets are comparable and fungible (that their units are interchangeable).

The UNFCCC has historically addressed carbon pricing in the context of the Kyoto Protocol's flexibility mechanisms, namely the CDM, Joint Implementation, and emissions trading between industrialized countries. After 2015, the focus has changed from the flexible mechanisms of the Kyoto Protocol to the specific operationalization of Article 6 of the Paris Agreement (UNFCCC 2015), which includes the linking of emissions-trading systems and offsets as well as non-market approaches. Within the UNFCCC as an institution, the international bureaucracy of

[13] Interview with senior official from the PMR, 27 August 2018.

the UNFCCC Secretariat[14] supports the negotiations and other activities of the UN climate regime, especially by providing information, arranging meetings, and drafting proposals (Busch 2009). Like other environmental regime secretariats, it performs key regime functions and has agency in its own right (Jinnah 2014, ch. 2). This said, the UNFCCC Secretariat has limited autonomy in its mandate and resources compared to the World Bank. The Secretariat is included in this analysis because of its important carbon-pricing activities, which can be divided into: (1) the support of negotiations specifying the contents of Article 6; (2) supporting the operation of clean development mechanisms; and (3) the supporting of countries that adopt carbon pricing to meet their Nationally Determined Contributions by providing technical advice, etc.[15]

Regarding the core norm of carbon pricing, the World Bank institutions are as such not permitted to promote official opinions about how carbon ideally should be priced, but they nonetheless have considerable autonomy from their member states (Nielson and Tierney 2005). The UNFCCC's position on carbon pricing reflects a compromise between its member states. Even more restricted than the World Bank institutions, the UNFCCC Secretariat is not permitted to hold an official position on how carbon ideally should be priced. Nonetheless, it is possible to identify how the UNFCCC Secretariat and the three World Bank institutions have addressed and framed carbon pricing in their day-to-day practices and, in this way, interpreted key aspects of the norm.

The World Bank as well as the institutions embedded within it have since 2014 stressed the importance of pricing carbon and, except for the NCM, have emphasized carbon taxes, emissions trading, and, to a lesser degree, offsets. The current framing aims to internalize the 'external costs of carbon emissions . . . and tie them to their sources through a price on carbon' and to 'shift the burden for the damage back to those who are responsible for it and who can reduce it. . . . In this way, the overall environmental goal is achieved in the most flexible and least-cost way to society' (World Bank 2018a). This framing and the bracketing of carbon taxes with carbon markets is a recent development. Prior to 2014, the World Bank focused on carbon markets and paid little attention to carbon taxes. Tellingly, the World Bank's influential annual report, which since 2014 has been named 'The State and Trends of Carbon Pricing', was from the initial publication in 2003 and until 2012 named 'The State and Trends of the Carbon Market' (no report was published in 2013).[16] This change is also visible in the content of the reports, with the pre-2014 reports focusing on the functioning of the carbon markets around the

[14] We use the term 'The UNFCCC' to refer to the institution as a whole, and state it explicitly when we refer to the Secretariat.

[15] Interview with UNFCCC Secretariat official, 3 July 2017.

[16] We are grateful to Matt Paterson for alerting us to this development.

globe and their total volume measured in tonnes of CO_2-equivalents and US dollars, rather than carbon pricing as an instrument to address the externality of climate change or shift the burden (World Bank 2012). Thus, the focus was put on creating functioning markets and linking them rather than pricing emissions and ensuring that those responsible pay for them. This position is close to that of other carbon-market institutions such as IETA. Finally, whereas the PMR (founded in 2010) and the NCM (founded in 2013) contain the word 'market' in their names, the CPLC (founded in 2014) focuses on carbon pricing.

The UNFCCC also changed focus, from initially concentrating on the Kyoto Protocol mechanisms to now targeting the mechanisms under Article 6 of the Paris Agreement as well as promoting carbon pricing as a policy instrument. This change was driven by the UNFCCC negotiation process that culminated in the Paris Agreement. The UN climate regime did not adopt specific positions on what domestic carbon pricing should look like, including whether polluters should pay for all of their emissions. Instead, it implicitly emphasized and facilitated particular practices, including a new offset mechanism, the Sustainable Development Mechanism, under the Paris Agreement, which focuses on sustainable development in a broader sense, rather than just mitigation. The UNFCCC Secretariat considered carbon pricing a key policy instrument, without defining it as a stand-alone instrument but rather as one among many.[17] Importantly, the Secretariat views carbon pricing as a tool to shift investment from carbon-intensive to sustainable means of production and to promote the deployment of the low-carbon technologies required for meeting the 2 or 1.5 degree target.[18]

6.4.2 Interlinkages

In terms of governance functions, all four institutions (UNFCCC, CPLC, PMR, and NCM) focus on sharing and creating new information as well as networking, while the PMR also engages in the provision of financing. These governance functions do not have the same inherent potential for conflict between the institutions as standard setting (see Chapter 2).

Regarding the core norm, the World Bank and its carbon-pricing institutions have tended to focus on first carbon markets and then carbon pricing as the crucial step in fighting climate change: once implemented, there is little reason to intervene politically in the subsequent causal chain leading to lower emissions. Nonetheless, there is also significant convergence between the UNFCCC and the World Bank institutions: they all place a strong emphasis on creating a functioning global

[17] Interview with senior UNFCCC official, 30 June 2017.
[18] Interview with UNFCCC Secretariat official, 3 July 2017.

carbon market, but in the interviews they did not take a stance on whether all emissions should be priced or if grandfathering was acceptable. Importantly, the World Bank has operated within the structures established by the UNFCCC (JI/CDM; NDCs; Article 6), but the main change in the perspective of the World Bank institutions (from a focus on carbon markets toward carbon pricing) did not originate in the UNFCCC. Likewise, all four institutions have promoted offsets as well as carbon pricing within given jurisdictions in various ways. The UNFCCC, the PMR, and the NCM all focus on mandatory carbon pricing, whereas the CPLC also has promoted voluntary carbon pricing within businesses.

A more important dividing line is the different confidence in the ability of the market. The World Bank institutions have been agnostic about how carbon pricing would lead to reduced emissions and framed the fact that carbon pricing leaves the decisions of how to mitigate to the market as a key strength. This neoclassical approach is based on the notion of the market as making the optimal choices. By contrast, the UNFCCC's approach to carbon pricing leaves considerable discretion to the states in the context of their Nationally Determined Contributions (NDCs) and only operates with carbon pricing as one instrument among many. Although the Paris Agreement and subsequent activities in the context of the Agreement's Article 6 endorse offset mechanisms and the linking of carbon markets, they very much leave any action up to the Parties and avoid talking about introducing a carbon price (Marcu 2016). The difference can be explained by the considerable autonomy of the World Bank institutions from their member states – compared to the UNFCCC set-up, in which states are involved in the decision-making process and can individually veto proposals.

6.4.3 Mechanisms

The UNFCCC and World Bank institutions interact in a range of different ways. First, and unlike for the subfield of carbon pricing in general, normative interlinkages play a major role. A sequence of rules from the UN climate regime, from the Kyoto Protocol to the 2015 Paris Agreement, have shaped much of the action of the World Bank institutions. The PMR is working with several countries to develop carbon-pricing policies that will help them achieving their NDCs. The NCM seeks to develop tools for linking of carbon markets that can be relevant under Article 6. Prior to the Paris Agreement, the Kyoto Protocol provided a similar context for the World Bank, which was key in developing JI/CDM – inter alia through its Prototype Carbon Fund and through its support for capacity building in countries seeking to host JI/CDM projects (Lazarowicz 2009; Lederer 2012). The World Bank institutions also promoted domestic carbon markets to help countries meet their targets under the Kyoto Protocol – and currently they (especially the PMR and the CPLC) promote carbon pricing as an instrument for countries to meet their NDC commitments.

Second, the World Bank institutions and the UNFCCC engage in behavioural interlinkages, especially through interlocking memberships. The UNFCCC Secretariat is an observer to the PMR and the CPLC, the same way that the World Bank is an observer to the UNFCCC. Importantly, several of the officials representing national governments within the PMR are also UNFCCC negotiators working on Article 6 within these negotiations.[19] Furthermore, The CPLC was launched at COP21 in Paris.

Third, the behavioural interlinkages often provided the basis for cognitive interlinkages, especially in terms of exchanging information and knowledge. The UNFCCC has especially collaborated with the PMR, both regarding turning the provisions of Article 6 into more concrete guidelines and providing support for countries adopting carbon pricing in the context of their NDCs.

6.5 Micro-Level Management

Drawing on the typology of micro-level management outlined in Chapter 2, it can be stated that the relationship between the UNFCCC and the World Bank–embedded institutions is managed jointly, with both sides trying to ensure compatibility between their activities. There were not any attempts of orchestration by third parties. Notwithstanding some unilateral low-key attempts, management was mainly bi- or multilateral and mainly took place through regular institutionalized contacts and meetings between officials. Officials working on carbon pricing constitute the main agents of management, whereas higher echelons of the World Bank and the UNFCCC (e.g. the World Bank Group Boards of Directors or the UNFCCC Executive Secretary) were less involved.

The institutions tend to collaborate in case they operate within the same countries, especially in Africa.[20] Regarding the support for countries adopting carbon pricing in the context of their NDCs, an informal division of labour has emerged bottom-up: the PMR mainly works with middle-income countries while the UNFCCC Secretariat concentrates on less developed countries.

Altogether, it makes sense to characterize the interlinkage between the UNFCCC and the World Bank institutions as one of coordination, although the management attempts have been taken in a bottom-up, incremental manner rather than as the result of overarching deliberate planning. The carbon-pricing subsystem has been constantly evolving and proliferating, which makes it more

[19] Interview with senior PMR official, 27 August 2018.
[20] Interview with senior PMR official, 27 August 2018.

difficult to assess the degree of coherence in a counterfactual no-management scenario. Yet, the informants interviewed for this study underscored the importance of management efforts in avoiding outright competition or conflict, albeit mainly for preventing deterioration. It is thus not possible to say whether the level of coherence has improved over time.

6.6 Conclusions

The analysis showed that the plethora of institutions that promote carbon pricing at the international level overlap to a significant degree in terms of geographical scope and governance functions. Most of them have global reach and membership, and several of them focus on either information-sharing and networking or standard-setting. The institutions were found to differ in terms of membership constituencies, with five public, three hybrid, and five private institutions. Mapping out institutional membership illustrated that the public and hybrid institutions are clustered around the World Bank, while the private ones circle around IETA. The institutions also differed in terms of their jurisdictional focus, with public and hybrid institutions mainly focusing on supporting political decisions to implement carbon pricing and to link carbon markets, and with private institutions focusing on the trading of emissions.

Altogether the field is characterized by coordination or coexistence, with significant attempts to establish a division of labour, and only little outright competition or conflict. Interlinkages have mainly been cognitive in nature (through workshops, co-developing knowledge, and information) and institutional (through interlocking memberships or written agreements). Generally, these interlinkages have been informal and took place between two or more institutions without significant differences in power.

We particularly focused on the interlinkage between, on the one hand, the UNFCCC and, on the other hand, the World Bank and the institutions embedded within it. The analysis showed that, despite the differences between the two camps, interlinkages are characterized by close coordination. This coordination has been mainly informed by a cognitive interaction mechanism – with institutions being observers at each other's meetings – and a normative mechanism – with the World Bank institutions operating within the framework set by the UNFCCC, particularly the offset mechanisms and the NDCs. Both the UNFCCC and the World Bank institutions promoted carbon pricing in general and a global carbon price in terms of linking carbon markets specifically. Differences between both sides were managed in a bottom-up, incremental fashion, which leads us to characterizing the relationship between the two institutions as one of coordination.

6.7 References

Abbott, K. W. 2012. The Transnational Regime Complex for Climate Change. *Environment & Planning C: Government & Policy* 30(4), 571–590.

Aldy, J. and Stavins, R. 2012. The Promise and Problems of Pricing Carbon. *Journal of Environment & Development* 21(2), 152–180.

Alter, K. J. and Meunier, S. 2009. The Politics of International Regime Complexity. *Perspectives on Politics* 7(1), 13–24.

Andersen, M. S. and Ekins, P. 2009. *Carbon-Energy Taxation: Lessons from Europe*. Oxford: Oxford University Press.

ATAG (Air Transport Action Group). 2016. Aviation Industry Views on CORSIA. Available at aviationbenefits.org/media/165859/Industry-position-paper_A39.pdf.

Ball, J. 2018. Why Carbon Pricing Isn't Working; Good Idea in Theory, Failing in Practice. *Foreign Affairs,* July/August 2018.

Benwell, R. 2009. Voluntary Aspects of Carbon Emissions Trading. *International Journal of Environmental Studies* 66(5), 605–618.

Bernstein, S. F. 2001. *The Compromise of Liberal Environmentalism*. New York: Columbia University Press.

Betsill, M. and Hoffmann, M. J. 2011. The Contours of 'Cap and Trade': The Evolution of Emissions Trading Systems for Greenhouse Gases. *Review of Policy Research* 28(1), 83–106.

Biermann, F., Pattberg, P., et al. 2009. The Fragmentation of Global Governance Architectures: A Framework for Analysis. *Global Environmental Politics* 9(4), 14–40.

Busch, P.-O. 2009. The Climate Secretariat: Making a Living in a Straitjacket. In *Managers of Global Change,* edited by F. Biermann and B. Siebenhüner. Cambridge, MA: MIT Press.

Choi, A. S. 2015. An Experimental Study to Explore WTP for Aviation Carbon Offsets: The Impact of a Carbon Tax on the Voluntary Action. *Journal of Environmental Planning & Management* 58(9), 1617–1634.

Coase, R. 1960. The Problem of Social Cost. *Journal of Law and Economics* 3(1), 1–44.

Felli, R. 2015. Environment, not Planning: The Neoliberal Depoliticisation of Environmental Policy by Means of Emissions Trading. *Environmental Politics* 24(5), 641–660.

Gallagher, K. P. and Yuan, F. 2017. Standardizing Sustainable Development: A Comparison of Development Banks in the Americas. *Journal of Environment and Development* 26(3), 243–271.

Harrison, K. 2010. The Comparative Politics of Carbon Taxation. *Annual Review of Law and Social Science*, 6, 507–529.

Harrison, K. 2012. A Tale of Two Taxes: The Fate of Environmental Tax Reform in Canada. *Review of Policy Research* 29(3), 383–407.

IATA (International Air Transport Association). 2009. IATA Press Release: First industry-wide program offers transparency and credible projects. Available at www.iata.org/pressroom/pr/Pages/2009-06-05-01.aspx.

ICAP (International Carbon Action Partnership). 2016. Emissions Trading in Practice: A Handbook on Design and Implementation. Available at https://openknowledge.worldbank.org/bitstream/handle/10986/23874/ETP.pdf?sequence=11&isAllowed=y.

IETA (International Emissions Trading Association). 2015. IETA's Views on the European Commission's Revision of the EU ETS Directive for the Post-2020 Period. Available at www.ieta.org/resources/EU/EU_2016/ETS%20revision%20Ph%204/IETA_position_paper_ETS_Revision_Final.pdf.

Jacobs, M. 1997. Sustainability and Markets: On the Neoclassical Model of Environmental Economics. *New Political Economy* 2(3), 365–385.

Jinnah, S. 2014. *Post-Treaty Politics*. Cambridge, MA: The MIT Press.

Kansy, T. 2016. Making the Links between Carbon Markets in a Post-Paris world. Available at blogs.worldbank.org/climatechange/print/making-links-between-carbon-markets-in-a-post-paris-world.

Keohane, R. O. and Victor, D. G. 2011. The Regime Complex for Climate Change. *Perspectives on Politics*, 9(1) 7–23.

Lane, R. and Newell, P. 2016. The Political Economy of Carbon Markets. In *The Palgrave Handbook of the International Political Economy of Energy*, edited by T. Van de Graaf, B. K. Sovacool, A. Ghosh, F. Kern, and M. T. Klare. London: Palgrave Macmillan.

Lazarowicz, M. 2009. *Global Carbon Trading – A Framework for Reducing Emissions*. London: UK Government.

Lederer, M. 2012. Market Making via Regulation: The Role of the State in Carbon Markets. *Regulation and Governance* 6(4), 524–544.

Lyon, T. P. and Maxwell, J. W. 2003. Self-Regulation, Taxation and Public Voluntary Environmental Agreements. *Journal of Public Economics*, 87, 1453–1486.

Lyon, T. P. and Maxwell, J. W. 2007. Environmental Public Voluntary Programs Reconsidered. *Policy Studies Journal* 35(4), 723–750.

Marcu, A. 2016. Carbon Market Provisions in the Paris Agreement (Article 6). CEPS Special Report: No 128. Brussells: Centre for European Policy Studies.

Meckling, J. 2011a. *Carbon Coalitions: Business, Climate Politics, and the Rise of Emissions Trading*. Cambridge, MA: MIT Press.

Meckling, J. 2011b. The Globalization of Carbon Trading: Transnational Business Coalitions in Climate Politics. *Global Environmental Politics* 11(2), 26–50.

Meckling, J. and Jenner, S. 2016. Varieties of Market-Based Policy: Instrument Choice in Climate Policy. *Environmental Politics* 25(5), 853–874.

NCM (Networked Carbon Markets). 2017. Networked Carbon Markets. Available at www.worldbank.org/en/topic/climatechange/brief/globally-networked-carbon-markets.

Newell, P. and Paterson, M. 2010. *Climate Capitalism: Global Warming and the Transformation of the Global Economy*. Cambridge: Cambridge University Press.

Nielson, D. L. and Tierney, M. J. 2005. Theory, Data, and Hypothesis Testing: World Bank Environmental Reform Redux. *International Organization* 59(3), 785–800.

Nordhaus, W. D. 2008. *A Question of Balance – Weighing the Options on Global Warming Policies*. New Haven, CT: Yale University Press.

OECD (Organization for Economic Co-operation and Development). 1974. Recommendation of the Council on the Implementation of the Polluter-Pays Principle. *OECD/Legal/0132*. Available at https://legalinstruments.oecd.org/en/instruments/11.

Ostrom, E. 1990. *Governing the Commons – The Evolution of Institutions for Collective Action*. Cambridge: Cambridge University Press.

Park, S. 2005. Norm Diffusion within International Organizations: A Case Study of the World Bank. *Journal of International Relations and Development* 8(2), 111–141.

Park, S. 2010. *World Bank Group interactions with Environmentalists: Changing International Organisation Identities*. Manchester, UK: Manchester University Press.

Paterson, M. 2012. Who and What are Carbon Markets for? Politics and the Development of Climate Policy. *Climate Policy (Earthscan)* 12(1), 82–97.

Paterson, M., Hoffmann, M., et al. 2014. The Micro Foundations of Policy Diffusion Toward Complex Global Governance: An Analysis of the Transnational Carbon Emission Trading Network. *Comparative Political Studies* 47(3), 420–449.

PMR (Partnership for Market Readiness). 2017. Supporting Action for Climate Change Mitigation, Available at www.thepmr.org/system/files/documents/PMRbooklet_1-2017.pdf.

Pigou, A. C. 1932. *The Economics of Welfare*. London: Macmillan.

Rabe, B. G. and Borick, C. P. 2012. Carbon Taxation and Policy Labeling: Experience from American States and Canadian Provinces. *Review of Policy Research* 29(3), 358–382.

Sanderink, L., Pattberg, P., et al. 2016. *Mapping the Institutional Architecture of the Climate-Energy Nexus*. Amsterdam: IVM Institute for Environmental Studies.

Segerson, K. 2013. Voluntary Approaches to Environmental Protection and Resource Management. *Annual Review of Resource Economics*, 5(1), 161–180.

Skjærseth, J. B. and Wettestad, J. 2008. *EU Emissions Trading – Initiation, Decision-making and Implementation*. Aldershot, UK: Ashgate.

Skovgaard, J., Sacks Ferrari, S., and Knaggård, Å. 2019. Mapping and Clustering the Adoption of Carbon Pricing Policies – What Polities Price Carbon and Why? *Climate Policy* 19(9), 1173–1185.

Stephan, B. and Paterson, M. 2012. The Politics of Carbon Markets. *Environmental Politics* 21(4), 545–562.

Sterner, T. and Coria, J. 2012. *Policy Instruments for Environmental and Natural Resource Management*. Washington, DC: RFF Press.

Stokke, O. S. 2001. *The Interplay of International Regimes: Putting Effectiveness Theory to Work*. FNI Report. Oslo: The Fridtjof Nansen Institute.

Tol, R. S. J. 2011. The Social Cost of Carbon. *Annual Review of Resource Economics* 3, 419–443.

UNFCCC (United Nations Framework Convention on Climate Change). 1992. United Nations Framework Convention on Climate Change.

UNFCCC (United Nations Framework Convention on Climate Change). 2015. Decision 1/CP.21 Paris Agreement.

VCS (Verified Carbon Standard). 2018. The VCS Program. Available at www.v-c-s.org/project/vcs-program.

World Bank. 2012. *The State and Trends of the Carbon Market 2012*. Washington, DC: World Bank.

World Bank. 2014. *State and Trends of Carbon Pricing 2014*. Washington, DC: The World Bank, Ecofys and Vivideconomics.

World Bank. 2018a. *Pricing Carbon*. Available at: www.worldbank.org/en/programs/pricing-carbon.

World Bank. 2018b. *State and Trends of Carbon Pricing 2018*. Washington, DC: The World Bank, Ecofys and Vivideconomics.

Part III

Legitimacy and Effectiveness in the Climate-Energy Nexus

7

Disentangling Legitimacy

Comparing Stakeholder Assessments of Five Key Climate and Energy Governance Institutions

NAGHMEH NASIRITOUSI AND SOETKIN VERHAEGEN

7.1 Introduction

This chapter builds on the insights of the previous chapters about the institutional complexity of the climate-energy nexus. As was shown in Chapter 3, the global institutional complex on climate and energy governance has in recent years developed into a crowded field with the emergence of several international institutions that seek to address both issues in tandem. Hence, multiple actors work in the same area without overarching coordination (Biermann et al. 2009). With different mandates, forms, functions, and values, these institutions both cooperate and compete with one another to further their mission. Given scarce resources amongst policy makers and other stakeholders, these actors need to prioritize which institutions to engage with.

Central to the question of which international institutions warrant support and are prioritized are considerations of the institutions' legitimacy. With competition over members and resources, international institutions depend on favourable perceptions of legitimacy by a diverse set of global governance stakeholders, such as policy makers, nongovernmental organizations, and businesses, to achieve their objectives (Andresen and Hey 2005; Biermann et al. 2009). As was discussed in Chapter 2, legitimacy broadly refers to 'the acceptance and justification of shared rule by a community' (Bernstein 2005, 142). Legitimacy is important for international institutions in order to be able to operate with authority and to attract constructive participation of political and societal stakeholders in the processes of making and implementing governance. Put differently, to achieve their objectives, international institutions must gain acceptance, trust, and credibility amongst the communities that they seek to govern (Andresen and Hey 2005).

The aim of this chapter is to understand how international institutions operating under institutional complexity are perceived by key stakeholders in terms of legitimacy. We present a novel approach to studying legitimacy perceptions as we capture stakeholders' assessments of a broad range of dimensions of legitimacy and bring those together in a composite measure of legitimacy assessments. Scholarly work on the concept of legitimacy highlights, and debates, that legitimacy is built on institutional qualities such as how the internal decision-making and accountability structures work and how effective and fair the institution is perceived to be (Scholte and Tallberg 2018). We contribute to this debate by showing that the surveyed stakeholders in climate and energy governance indeed perceive these elements as dimensions of the broader concept of legitimacy.

Concretely, by focusing on those aspects of legitimacy that international institutions themselves can influence, i.e. their institutional qualities, we contribute to understanding how perceptions of these – i.e. what we call legitimacy assessments – differ between stakeholder groups. Previous literature has to our knowledge not mapped stakeholder's perceptions of a set of institutions that work on similar issues and that thereby have overlapping mandates. In terms of empirical novelty, the chapter offers a systematic and comparative mapping of stakeholders' legitimacy assessments of five institutions. To this end, it uses a hybrid approach focusing on stakeholders' assessments of those dimensions of legitimacy that concern institutional qualities. Theoretically, the chapter unpacks the meaning of legitimacy under institutional complexity.

This chapter thereby provides innovative insights to the literatures on both legitimacy and institutional complexity, with implications for ways in which climate and energy governance can be strengthened. Moreover, the findings have implications for how institutions may influence perceived legitimacy deficits through legitimation strategies toward different stakeholder audiences (Bäckstrand and Söderbaum 2018).

We gained insight into stakeholders' legitimacy assessments by fielding an expert survey among energy and climate stakeholders from different world regions. Respondents were asked about five climate and energy governance institutions that exhibit different but overlapping mandates and membership: the Clean Energy Ministerial (CEM), the International Energy Agency (IEA), the International Renewable Energy Agency (IRENA), the Renewable Energy Policy Network for the 21st Century (REN 21), and the United Nations Framework Convention on Climate Change (UNFCCC). All five institutions belong to the subfield of renewable energy. As shown in Chapter 4, they play a key role for this subfield. Based on that chapter and its analysis of institutional coherence and management on renewable energy, we now expand the analysis of this subfield toward institutional legitimacy. The stakeholders who evaluate the five selected institutions comprise both state and nonstate actors, work with different issues (e.g. mitigation, adaptation, energy security, climate finance, and technology) and come from varying regions of the world. The data allow us to show how legitimacy assessments of these five institutions differ across stakeholder types and across stakeholders working with different issues.

The chapter proceeds as follows. The next section presents our framework for studying stakeholders' legitimacy assessments. Here, we also further develop the conceptual insights on legitimacy introduced in Chapter 2. Next, the five institutions in climate and energy governance are described, paying specific attention to the institutional qualities that are expected to be relevant in guiding legitimacy assessments. Thereafter, the data and methods section outlines how we measured these assessments. The results section then maps stakeholders' legitimacy assessments of the five institutions in our study. The final section summarizes the insights gained and highlights avenues for further research.

7.2 Theory and Concepts

As argued in Chapter 2, institutional complexity complicates an evaluation of legitimacy of individual institutions because of the interlinkages and overlapping mandates between institutions. In this section, we link back to the discussion in Chapter 2 on the concept of legitimacy and highlight how the cognitive model of legitimacy provides insights into understanding legitimacy under institutional complexity. Thereafter, we discuss the institutional qualities that have been argued to be central to institutions' legitimacy, deriving nine dimensions of legitimacy.

7.2.1 Congruence and Cognition: Understanding Perceptions of Legitimacy

The traditional view of legitimacy in IR has held that 'legitimacy depends on the congruence between an organization's features – specifically, its procedures, purpose, and performance – on the one hand, and the inter-subjectively shared norms and values held by relevant organizational stakeholders, on the other hand' (Lenz and Viola 2017, 943). Legitimacy in this view depends on the extent to which an institution lives up to certain legitimacy demands that stakeholders have, which are determined by the norms and values of those stakeholders. Recent research by Lenz and Viola (2017) has, however, outlined several empirical and analytical weaknesses in the traditional approach – or what they call 'the congruence model of legitimacy'. Central to this argument are limitations to stakeholders' ability to make a precise and complete evaluation of an institution in order to compare this to their normative beliefs.

Instead, Lenz and Viola (2017) introduce a 'cognitive model' for understanding how legitimacy perceptions are formed. This model draws on the literature on cognitive psychology to outline the micro-foundations for understanding the formation of legitimacy perceptions and reflects similar approaches in the public opinion literature (Armingeon and Ceka 2014). The three core insights that inform their model are: '(1) judgments rely on cognitive schemata and heuristics that bias judgments; (2) they are comparative; and (3) they are sticky, up to a threshold' (Lenz and Viola 2017, 947–948).

According to these insights, legitimacy perceptions are not formed in a vacuum, i.e. actors do not judge institutions one by one against their held social values and norms. Rather, perceptions of an institution are based on a reference point that is derived from previous experiences. These heuristics consist of perceptions of institutions that stakeholders are most familiar with or which they most recently engaged with, but it may also consist of an ideational prototype of what the perfect institution would look like. Heuristics are presented as rather stable images in stakeholders' minds. When we ask stakeholders to assess the legitimacy of an institution, we should therefore expect them to compare the perceived qualities of that institution to those of their 'heuristic' institution. Moreover, we can expect variations across stakeholders as they will have different reference points, or heuristics, depending on their background, the institutions they are mostly familiar with, and the norms they hold.

While this turns legitimacy assessments into something much more personal than the congruence model proposes, processes of socialization and shared experiences within specific professional sectors lead us to expect systematic similarities in the used heuristics and normative beliefs about legitimacy across individuals within the same sector, and differences among individuals in different sectors. For instance, nonstate actors such as business or civil society actors may assess institutions in relation to the norms of legitimate governance that are central in their respective peer group. Likewise, climate- and energy-related stakeholders that also work on questions of international development are expected to keep development institutions, and their respective norms, in mind when they assess the legitimacy of the climate and energy governance institutions in our study. This very use of heuristics, as well as its dependence upon stakeholders' specific experiences, provides an additional motivation for studying individual legitimacy assessments (Scholte and Tallberg 2018).

This conception of legitimacy has two key implications for how we can understand legitimacy perceptions. First, this chapter argues that an awareness of cognitive limitations is central to understanding legitimacy beliefs. Rather than assuming that actors, even if they are experts, are capable of capturing the exact way in which institutions function and the extent to which the institution is in line with those actors' normative beliefs, one should recognize that legitimacy assessments are based on heuristics and underlying experiences, which come with respective limitations. Especially in a highly complex, and therefore cognitively demanding, institutional environment, one may expect actors to base their legitimacy assessments on such heuristic simplifications. When facing several institutions with overlapping and complex mandates, actors may use mental shortcuts to form opinions about some of these institutions (Alter and Meunier 2009).

Second, the norms, values, and experience of actors can both influence how they assess the qualities of an institution as well as how they value these qualities, i.e. the relative importance that they place on the purpose, process, or performance of institutions. In other words, stakeholders' legitimacy perceptions may differ either because they assess the institutional qualities of institutions differently, and/or because they value different characteristics of legitimacy differently. This means that an actor's legitimacy perceptions, i.e. the extent to which an institution is viewed as legitimate by an actor, is a combination of that actor's legitimacy assessment (i.e. an assessment of the institutional qualities of an institution) and that actor's legitimacy valuation (i.e. the importance attached to certain institutional qualities). This chapter focuses on legitimacy assessments by climate and energy experts along nine dimensions of legitimacy as explained in the next section.

7.2.2 Legitimacy Criteria Used to Map Perceptions

Legitimacy is the assessment and valuation by an audience as to the appropriateness of an authority. What *should* be considered a legitimate form of authority has preoccupied normative scholars. What *is in practice* considered a legitimate form of authority is instead the focus of sociological work (Nasiritousi et al. 2016). In this chapter we opt for a hybrid approach, as we study stakeholders' perceptions of institutions while referring to normative criteria of legitimate governance (cf. Agné 2018). This take thus differs from a 'purely' sociological approach where it is left to selected stakeholders to determine relevant criteria for assessing an institutions' legitimacy. In this type of study, legitimacy is empirically measured as confidence in, or support for, an institution (Gibson and Caldeira 1998; Dellmuth and Tallberg 2015).

The current study, in contrast, combines normative and sociological aspects. It does so by seeking to understand legitimacy in terms of its different dimensions. This approach provides a uniquely fine-grained perspective on legitimacy perceptions (cf. Scholte and Tallberg 2018). The hybrid approach is in line with the work of Beetham (1991), who argues that legitimacy has both a normative and sociological component, as perceptions of institutions' legitimacy will be based on institutions meeting normative criteria on the exercise of power.

Concretely we seek to provide a comparative mapping of stakeholders' views of a set of nine institutional qualities or dimensions derived from the normative literature. This helps us to better understand how legitimacy assessments may vary between different institutions and various stakeholder groups. These assessments are expected to be an important indicator for sociological legitimacy (cf. Scholte and Tallberg 2018).

Our conceptual framework therefore begins with identifying dimensions of legitimacy. We do so by advancing normative criteria, i.e. a set of standards that are 'grounded in normative theories that reflect prevailing sociological standards in society' (Karlsson-Vinkhuyzen and McGee 2013, 58). Central to the identification of dimensions of legitimacy is the distinction between input and output legitimacy. While input legitimacy refers to the design of political processes, i.e. governance *by* the people, output legitimacy concerns problem-solving capacity, i.e. governance *for* the people (Scharpf 1999). By exploring aspects of input and output legitimacy, it is possible to derive criteria for assessing legitimacy anchored in a normative framework.

The normative framework presented in Table 7.1 builds on the works of Bodansky (1999), Karlsson-Vinkhuyzen and Vihma (2009) and Mena and Palazzo (2012). The framework distinguishes source-based and process-based input legitimacy, as well as substantial and distributive output legitimacy. These unfold into a total of nine dimensions of legitimacy.

Source-based legitimacy refers to how authority is gained by an institution – not by its operations, but through its essence and standing. Three common forms of source-based legitimacy are expertise, tradition, and discourse (Karlsson-Vinkuyzen and McGee 2013). **Process-based legitimacy** pertains to the design of procedural rules that affect the decision making of the institution. *Inclusion* refers to how open the institution is in terms of membership. *Procedural fairness* in decision making means that stakeholders have opportunities to be heard and be treated fairly so as to have a sense of ownership of the decisions made (Raines 2003). *Transparency* relates to the degree of access to information that the institution provides to members and other stakeholders. *Accountability* implies that institutions can be held to account for the decisions that they make and for the ways in which they implement these decisions. **Substantial legitimacy** is concerned with issues of effectiveness. *Output* concerns performance in terms of what the international institution produces, for example issuing regulations (these can be binding or non-binding), producing reports, conducting research, organizing meetings, providing funding, providing training, etc. (Szulecki et al. 2011). *Outcome* relates to whether the institution produces behavioural changes, for example in terms of whether the institution increases the level of cooperation and compliance amongst members for instance by improving learning and modifying incentives (Underdal 2002; Gutner and Thompson 2010). To determine an institution's *impact* involves making judgements about the extent to which the institution

Table 7.1 *Analytical Framework - Dimensions of Input and Output Legitimacy and their Operationalization.*

Input or Output Legitimacy	Dimensions of Legitimacy	Operationalization in Survey
		For those institutions in Question 6 that you are familiar with (where you answered 3–5), please evaluate these institutions in their respective column according to the criteria below. Write a score between 1–5 in each cell, where 1 means that the institution is very weak and 5 means it is very strong on the respective dimension.
Source-based (input) legitimacy	Source of authority	Expertise
Process-based (input) legitimacy	Inclusion	Inclusion of all appropriate actors
	Procedural fairness	Procedural (decision-making) fairness
	Transparency	Transparency
	Accountability	Accountability
Substantial (output) legitimacy	Output	Output (what is produced)
	Outcome	Outcome (the effect the output has on its members)
	Impact	Impact (the effect the output has on problem-solving)
Distributive (output) legitimacy	Distributive fairness	Distributive fairness (distributing benefits to members fairly)

Source: CLIMENGO Expert Survey 2017–2018.

contributes to alleviating the problem it was tasked to resolve (Underdal 2002). Distributive legitimacy, finally, is a dimension that is concerned with the distribution of benefits to the members of the institution.

7.3 The Five Cases: Similarities and Differences in Institutional Qualities

The five institutions whose legitimacy we put under scrutiny in this chapter are: CEM, IEA, IRENA, REN21, and UNFCCC. These institutions have different forms and functions, yet they also have overlapping mandates. We selected these institutions since they pertain to one major subfield of the climate-energy nexus, namely renewable energy. Chapter 4 analyzed the degree of coherence of the renewable energy subfield and identified these as the key institutions therein (Sanderink, this volume). Their importance was further confirmed by climate and energy experts (both state and nonstate actors) that we interviewed prior to designing our questionnaire. The five institutions have thus all achieved a certain level of authority, which makes them interesting cases for a comparative mapping of how stakeholders' legitimacy assessments differ amongst these key institutions. In what follows, we briefly introduce the five institutions based on their self-descriptions – by representatives we approached or on their websites[1] – and highlight a number of similarities and differences across them in terms of key properties. The descriptions form the context for our expectations that we thereafter derive about how stakeholders make legitimacy assessments.

The most long-standing institution in our sample is the IEA – an intergovernmental organization that was established in 1974 and is based in Paris. The IEA was established within the framework of the Organisation for Economic Co-operation and Development (OECD) in response to the 1973 oil crisis to strengthen the cooperation of industrialized countries to meet the energy needs of oil-consuming countries. The agency draws its thirty member countries from the OECD group of industrialized countries, and, in addition, features eight association countries: Brazil, China, India, Indonesia, Morocco, Singapore, South Africa, and Thailand (IEA 2018a). Association countries may participate in the analytical work of the IEA, but have no rights and obligations. While its main focus has been to tackle global oil supply disruptions, the IEA's mandate has broadened to 'ensure reliable, affordable and clean energy for its thirty member countries and beyond' (IEA 2018b). It has a global scope and works on energy security, sustainability and clean energy transitions, technology, innovation, and energy access. The main decision-making body of the IEA is the Governing Board, which comprises energy ministers or their senior representatives from each member country. Governing Board decisions are legally binding on all member countries. Majority vote is based on a system of voting weights allocated to each member country. Such a vote is required for decisions on the IEA Programme of Work, procedural questions, and recommendations. Unanimity is required for other decisions. The IEA works closely with partners, including industry partners, and other international institutions to gain insights and advice from outside actors (IEA 2018c). There is no formal role for nonstate actors, but nonstate actors may contribute to and peer-review IEA reports, participate in IEA events and programmes, and serve on IEA advisory boards. In terms of output, the IEA collects data, conducts research, provides analysis, makes policy recommendations, produces reports, organizes meetings/workshops/seminars, and offers training.

[1] The UNFCCC representatives we contacted declined to answer our specific questions and instead referred us to the information on their website.

Almost two decades after the establishment of the IEA, countries adopted the United Nations Framework Convention on Climate Change in 1992. With near-universal membership, the objective of the UNFCCC is to 'stabilize greenhouse gas concentrations in the atmosphere at a level that would prevent dangerous anthropogenic interference with the climate system' (UNFCCC 2018). Tasked with supporting the operation of this international environmental treaty, the UNFCCC Secretariat is based in Bonn. The UNFCCC is an intergovernmental institution that makes decisions based on consensus. The UNFCCC deals with a range of issues related to climate change, including mitigation, adaptation, technology, capacity building, and finance. It is also one of the most open international institutions in terms of involving a range of nonstate actors in the yearly conferences compared to other institutions in, for example, trade or security fields (Nasiritousi and Linnér 2016). Nonstate actors also have a prominent role in the Global Action Agenda, an initiative to spur more ambitious climate action amongst stakeholders, as evidenced by the Yearbook for Global Climate Action (UN Climate Change Secretariat 2018) and the NAZCA database of climate commitments by nonstate actors. The UNFCCC's key outputs have been the 1997 Kyoto Protocol and the 2015 Paris Agreement, both landmark international agreements aimed at addressing the causes and consequences of climate change.

More recent institutions are REN21, IRENA, and CEM. REN21 was launched in 2004 as a 'global renewable energy policy multi-stakeholder network' (REN21 2018a). It is based at the office of the UN Environment Programme (UNEP) in Paris. Its mission is to facilitate knowledge exchange and drive a transition toward renewable energy. The members of REN21 come from five stakeholder groups: governments, industry associations, nongovernmental organizations (NGOs), academia, and other international organizations. REN21 tries to keep membership balanced between the five stakeholder groups. By implication, governments are outnumbered by nonstate actors. Government representatives come from the following thirteen countries: Afghanistan, Brazil, Denmark, Dominican Republic, Germany, India, Mexico, Norway, Republic of Korea, South Africa, Spain, United Arab Emirates, and the United States (REN21 2020). REN21 is thus a collaborative network that seeks to connect the public and private sectors on renewable energy (REN21 2018b). The Steering Committee is elected from REN21's members, ten from each stakeholder group. From that, the seven people of the Bureau are elected. These elections are held at the annual meeting, the General Assembly, and this is the only time REN21 takes decisions by majority vote. Other decisions are typically consensus based. The Bureau provides month-to-month oversight while the Steering Committee conducts the broader, programmatic oversight. REN21's key output is the annual Global Status Report, which presents a rich set of data on the status of renewables and is widely disseminated among actors in the field.

Founded in 2009, IRENA is an intergovernmental organization that is headquartered in Abu Dhabi. It currently has 161 member states, with further 22 states currently undergoing accession processes (IRENA 2020a). The agency seeks to promote adoption and sustainable use of all forms of renewable energy, in the pursuit of sustainable development, energy access, energy security and low-carbon economic growth. The main decision-making body of IRENA is the Assembly, which includes one representative from each member country. All matters of substance are decided by consensus among the members present, whereas questions of procedure are decided by simple majority. IRENA works with the broader renewable energy community, including companies, NGOs, and other international organizations, to facilitate knowledge-sharing (IRENA 2020b). Examples include a joint project facility, online info and marketplace platforms, initiatives, and the Coalition for Action (IRENA 2020c). In terms of output, IRENA is involved in many activities, including: research and publication of reports, providing member states and nonstate actors with recommendations, issuing non-binding regulations, and providing training and funding to support implementation.

Established in 2009, CEM is a high-level ministerial forum that seeks to advance clean energy technologies by promoting initiatives based on common interests among its members and other stakeholders. Its Secretariat is seated at the IEA headquarters in Paris. CEM members include twenty-seven country governments, but also the European Commission. It is the only regular meeting of ministers focusing on clean energy. Rather than relying on consensus, CEM employs a 'distributed leadership' model whereby any government interested in furthering an idea on clean energy technology is encouraged to identify willing partners and proceed. The initiatives, which countries join based on their interests and capabilities, must include three or more CEM members, be endowed with resources, and offer a tangible work plan. CEM's work is divided into three general work categories: (1) energy supply systems and integration, (2) energy demand, and (3) cross-cutting support. The latter includes, for example, initiatives such as Women in Clean Energy and the Clean Energy Solutions Centre (which provides policy toolkits). In terms of output, each initiative sets its own deliverables and objectives, depending on their goals. Some produce reports and analysis, others focus on policy solutions, yet others use workshops, seminars, webinars, and other forms of knowledge-sharing. CEM also seeks the input of key private sector partners through, for instance, dedicated actions, commitments, or the hosting of workshops (CEM 2018).

These five institutions thus all operate in the complex of institutions that govern the climate-energy nexus within the subfield of renewable energy, but differ in a number of respects that may impact on how stakeholders assess their legitimacy. The first is in membership, where some are intergovernmental organizations with near-universal membership (UNFCCC, IRENA) while others are minilateral institutions (IEA and CEM) or multi-stakeholder partnerships (REN21). This may have implications for stakeholders' assessments of their inclusion, procedural fairness, and distributive fairness. Second, they differ in terms of the scope of their mandate, where some have a broad mandate focusing on multiple issues (UNFCCC and IEA) whereas others concentrate on more specific questions (IRENA, REN21, and CEM). Third, they vary in terms of the nature of their mandate, with the UNFCCC having a political mandate requiring negotiations on contentious issues between countries, while the other four institutions in the sample are endowed with a more technical mandate focusing on implementation. Their mandate is likely to have implications for stakeholders' assessments of the output, outcome, and impact of respective institutions. Fourth, the selected institutions differ with respect to how strongly they work with nonstate actors. The UNFCCC and REN21 have a close relationship with a broad range of nonstate actors in terms of access or cooperation. Other institutions are less engaged with such actors or are more selective, with a narrower set of nonstate collaboration partners (IRENA, CEM, IEA). This, in turn, may well affect stakeholders' assessments of their levels of inclusion and expertise. Fifth and finally, most institutions take decisions of substance based on consensus, whereas CEM has a more flexible decision-making structure where initiatives only need agreement between at least three members. This may have consequences for how stakeholders view procedural fairness and distributive fairness.

7.4 Theory-Based Expectations of Legitimacy Assessments

Some of the differences mentioned in the previous section have theoretical value, since they imply expectations about legitimacy assessments. In what follows, we turn to the question of how stakeholders' assessments of the legitimacy of the five key institutions governing the climate-energy nexus may vary.

The literature has shown that different types of stakeholders hold different legitimacy demands based on their social values, norms, and previous experiences (Bernstein 2005; Karlsson-Vinkhuyzen and Vihma 2009; Lenz and Viola 2017). We argue that legitimacy demands can therefore vary depending on (1) the type of stakeholder (i.e. government, business, or NGO representative); (2) the issues that these stakeholders primarily work with (for example energy, development, or climate change); and (3) where in the world the person comes from, as social values, norms, and experiences can be expected to vary across different legitimacy-granting communities (Symons 2011; Nasiritousi et al. 2016). Thus, stakeholder type, focus of work and geographical origin can serve as proxies for differences in norms, values, and experiences that may influence legitimacy assessments.

At the same time, institutional complexity – and the logic of the cognitive model – implies that stakeholders, even if they are experts, may face difficulties in distinguishing their assessments of institutions that are similar in their functions due to bounded rationality (Alter and Meunier 2009). If it is indeed too hard for stakeholders to disentangle certain properties across institutions with overlapping mandates, e.g. dimensions such as outcome and impact (Bäckstrand et al. 2018), the cognitive model of legitimacy would lead us to expect that there will not be great variation in stakeholders' assessments of the institutions governing the climate-energy nexus. Despite the differences in institutional qualities outlined in the previous section, the five institutions are interrelated and fulfil comparable governance functions within the same subfield, such that stakeholders might draw on similar heuristics to form their legitimacy assessments. In sum, the literature provides reasons to expect both variation and similarity in legitimacy assessments of institutions, across different categories of stakeholders.

Expectations can therefore be drawn up based on the nature of institutions as well as on the background of the stakeholders. The following expectations will guide the exploratory analysis that we present in the remainder of this chapter. First, all five institutions are relatively specialized and rely on expert knowledge as source-based input legitimacy. It is therefore of interest to explore whether stakeholders agree with the institutions' claims that they are strong on expertise. Given that expertise is an important feature of the institutions studied, we have reasons to believe that the expertise dimension will be positively evaluated by stakeholders. Conversely, because most institutions are more concerned with expertise than the empowerment of marginalized groups, procedural and distributional fairness can be expected to be evaluated more negatively (cf. Nasiritousi et al. 2016).

Second, the selected institutions vary along the nine dimensions of legitimacy. Particularly the UNFCCC fulfils many of the respective normative criteria, with, for example, inclusive membership, relative openness toward nonstate actors, and outputs such as the Paris Agreement and can therefore be expected to rank highly on legitimacy (Karlsson-Vinkhuyzen and McGee 2013). Yet, the cognitive model highlights that legitimacy assessments also depend on the prototype used by actors to form their perceptions (Lenz and Viola 2017). This implies that, while an institution fulfils many normative criteria of legitimacy, legitimacy assessments may still vary depending on the norms, values and experiences of the community of stakeholders that grant legitimacy.

Third, and linking to the background of stakeholders, we may expect different legitimacy assessments among state actors on the one hand, and nonstate actors on the other. State actors play an important role in intergovernmental organizations, and are likely to take these as a point of reference. For nonstate actors, on the other hand, the prototype used to make an evaluation is likely to be an institution that the nonstate actor is familiar with or wishes for, i.e. a relatively open institution with formal access for nonstate actor participation (Tallberg et al. 2014). Institutions that are relatively closed are therefore more likely to be negatively evaluated by nonstate actors than by government representatives.[2]

Fourth, stakeholders also differ in terms of the issue areas they are predominantly working on. Differences in legitimacy assessments could thus also arise from variations in norms and values that go back to different thematic environments. Stakeholders from a certain community are likely to be more familiar with institutions from their own field than from other issue areas and, subsequently, may well use different heuristics or prototypes. For example, those actors working primarily in the energy sector may be much more familiar with institutions such as the Organization of the Petroleum Exporting Countries (OPEC) and the International Energy Forum than those actors that primarily work in the development sector – who, in turn, may be more familiar with, for example, the Global Environment Facility and the Green Climate Fund. In other words, the frame of reference that actors in global climate and energy governance use for their legitimacy assessments can be assumed to reach far beyond the institutions included in this study.

Finally, we expect to observe differences in legitimacy assessments based on where respondents come from. Both legitimacy norms and heuristics are likely to vary depending on the geographical background of the respondents. For instance, governance norms and expectations, political culture, and level of involvement in international organizations may differ considerably across countries. This said, our sample consists of experts largely active in international circles. This might weaken the differentiating effect of geographical origin as these experts may have experienced a certain socialization into more general and international norms of global governance (Flockhart 2006; Greenhill 2010).

In what follows, we use our expectations as an explorative guidance to provide a first empirical mapping of legitimacy assessments for the five selected key institutions governing the climate-energy nexus. This mapping will offer novel insights into how these assessments differ between institutions and stakeholders.

7.5 Data and Methods

This chapter uses unique questionnaire data to capture the assessments of key stakeholders on the different dimensions of input and output legitimacy that we

2 This said, a rational/utilitarian logic could also be at play here, e.g. when nonstate actors favour institutions that give them influence (Verhaegen et al. 2018). In this chapter, however, and as outlined in our analytical framework, we approach legitimacy as going beyond motives of self-interest (Agné 2018).

introduced previously (CLIMENGO Expert Survey 2017–2018). Climate and energy experts were surveyed, including representatives from national, regional, and local governments as well as businesses, NGOs, academia, and intergovernmental organizations. The survey was distributed to participants at three venues: the UNFCCC COP23 in Bonn, Germany, November 2017; the UNFCCC intersessional in Bonn, Germany, May 2018; and the Nordic Clean Energy Week that comprised both Mission Innovation and CEM meetings in Malmö, Sweden, and Copenhagen, Denmark, May 2018. At the UNFCCC meetings we handed out questionnaires in side-events with an energy-related focus. We thereby obtained responses from a broad range of public and private stakeholders that work with climate and energy questions.

In addition, we created an online version of the questionnaire to target specific categories of respondents that were not sufficiently covered by the paper version of the survey. As probability sampling was not possible – given that it is not possible to define the population of climate and energy experts in global governance – we aimed at covering a broad variety of stakeholders. This means that, while we can show differences in legitimacy assessments between stakeholder categories, the results cannot be extrapolated to the entire population of climate and energy experts in global governance.

The survey first asked respondents to indicate which type of stakeholder they are, and which issue areas are central to their work. They were also asked to indicate their nationality. Next, respondents were asked how familiar they are with the five institutions of our study. When respondents indicated to be at least somewhat familiar with an institution, they were asked follow-up questions on nine criteria that reflect the different dimensions of legitimacy as identified in the conceptual framework. Respondents were instructed to use a scale that ranges between 1 (very weak) and 5 (very strong) to evaluate each organizations' expertise, transparency, accountability, inclusion of all appropriate actors, procedural (decision-making) fairness, output (what is produced), outcome (the effect the output has on its members), impact (the effect the outcome has on problem-solving), and distributive fairness (distributing benefits to members fairly).

The survey was completed by 262 respondents in total. Of these, 28 per cent were government representatives, 26 per cent represented an NGO, 23 per cent identified themselves as academics, 17 per cent represented a business organization, and 8 per cent an intergovernmental organization. The largest share of respondents worked with multiple issue areas; most of them with climate mitigation (36 per cent), followed by technology (30 per cent), energy or energy security (30 per cent), development (19 per cent), adaptation (19 per cent), and climate finance and carbon pricing, e.g. carbon markets (17 per cent). Geographically, most respondents hold a European or Western nationality[3] (64 per cent); 13 per cent of respondents came from Africa, 15 per cent came from the Asia-Pacific region, and 6 per cent from a Latin American or Caribbean country.

These groupings were used to examine differences in perceptions of stakeholders from different geographical origins. An additional categorization of nationalities was conducted based on the World Bank's income categories of countries, i.e. low, lower-middle, upper-middle, and high. This two-pronged approach allows us to test whether differences in legitimacy assessments stem from differences in norms, values, or experiences held across world regions as determined by geography or by income.

As respondents could indicate multiple actor types and issue areas they were active in, t-tests (rather than analysis of variance, i.e. ANOVA) were performed in order to explore statistical differences in their legitimacy assessments. For each instance, all respondents who indicated to be active within a certain actor type or to be working with a certain issue area were compared to all respondents who were active in a specific other organization, or active in another issue area. Among the surveyed stakeholders, the most well-known organization is the UNFCCC, with 90 per cent of respondents being rather to highly familiar with this organization. Next are the IEA (87 per cent), IRENA (84 per cent), and REN21 (47 per cent), while only 35 per cent of the respondents are at least rather familiar with CEM.[4]

7.6 Results

7.6.1 Exploring Nine Dimensions of Legitimacy

Figure 1 presents assessments by the respondents for each institution and each legitimacy dimension, as well as the average score for all dimensions taken together ('total average' in the figure). On the whole, we see that expertise is positively assessed across the institutions. For the IEA, IRENA, REN21, and UNFCCC it ranges between 4 and 4.5 points on the 5-point scale. Specifically, the data show that the level of expertise of the IEA and IRENA is significantly more positively evaluated by the respondents than any other legitimacy dimension of those institutions (confirmed by a t-test). Similarly, for REN21 and the UNFCCC the level of expertise is more positively evaluated than most other

[3] Follows the UN Regional Groups but also includes Eastern European countries, see: www.un.org/depts/DGACM/RegionalGroups.shtml.

[4] Given that two out of three venues at which surveys were distributed were UNFCCC meetings, the high familiarity with the UNFCCC is not very surprising. Yet, we also distributed surveys at the Nordic Clean Energy Week in 2018, which also comprised a CEM meeting. The overall results thus particularly demonstrate that even among experts in climate and energy governance, CEM is not very well known.

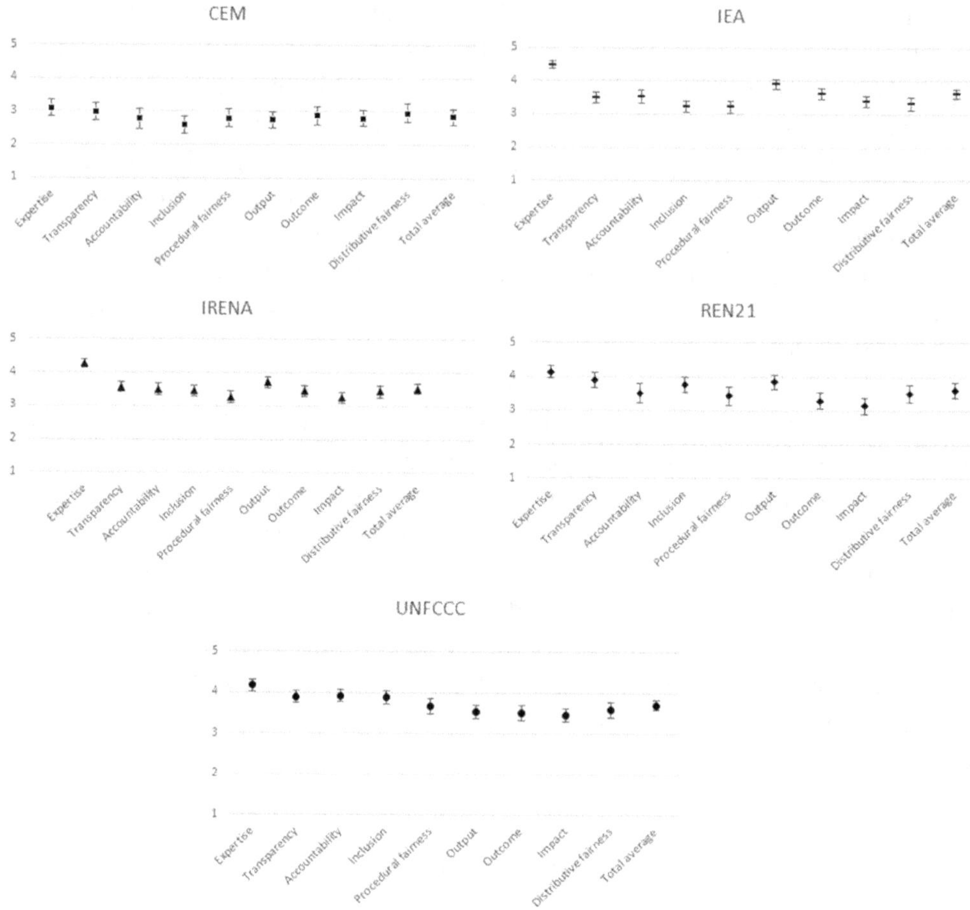

Figure 7.1 Mean levels of legitimacy assessments per institution.

dimensions. This is in line with our expectations, as this dimension represents a key feature of the institutions we study and the survey shows that it is recognized accordingly by the respondents.

Additionally, we observe that for the UNFCCC, evaluations of its input legitimacy are on average more positive than those of its output legitimacy. For the other institutions, however, no such division is visible. Furthermore, the evaluation of the input legitimacy of the UNFCCC is higher than that of CEM, IEA, and IRENA, while the perceived output legitimacy of the UNFCCC is similar to that of the IEA, IRENA, and REN21. One possible explanation for the UNFCCC's strong performance on input dimensions could be that the highly political negotiations have forced the institution to put increased emphasis on strengthening inclusion and transparency to maintain legitimacy, at least in comparison to the other institutions in our study. This has particularly been highlighted in the aftermath of the Copenhagen conference in 2009 (Karlsson-Vinkhuyzen and McGee 2013). Moreover, transparency and inclusion were key to the French Presidency that was successful in concluding the Paris Agreement (Brun 2016).

7.6.2 Exploratory Factor Analysis of Legitimacy Dimensions

The next step in the analysis uses exploratory factor analysis in order to examine the underlying structure in the data. Exploratory factor analysis is a statistical method that is used to determine how many distinct constructs are captured by a set of measures, in a case where the researcher does not have definite expectations about the underlying structure of correlations between the observed measures. In other words, it shows whether the measures capture different aspects of one broader construct or whether they capture multiple constructs (Fabrigar and Wegener 2012).

In this particular study, we test whether the nine dimensions of legitimacy together measure one underlying construct, 'legitimacy assessments', or whether they load on two separate factors, as they might as well capture 'input legitimacy assessments' on the one hand, and 'output legitimacy assessments' on the other. The factor loadings and Eigenvalues of the exploratory factor analyses for each institution indicate that the items indeed load on one underlying factor which we label 'legitimacy assessments'. Hence, the individual assessments of each legitimacy dimension can be treated as part of a broader measure for legitimacy assessments, in a multi-faceted manner.

For this reason, the composite indicator (a sum-scale ranging from 1 to 5) for perceived legitimacy of each institution was used in the remainder of the study as a measure for respondents' legitimacy assessments. On average, respondents' assessment of CEM across the different dimensions of legitimacy (mean=2.845) is less positive than that of any of the other institutions, while the legitimacy assessment of the UNFCCC (mean = 3.707) is the most positive for all institutions in our study (confirmed by t-tests). This suggests that, in comparative terms, the UNFCCC is perceived to meet best the normative expectations of respondents – which corresponds to the institution's relatively good formal record on some of the criteria. The average overall legitimacy assessments of the IEA, IRENA, and REN21 do not significantly differ from one another. Thus, this first observation indicates that the extent to which institutions formally meet normative legitimacy criteria has an influence on individual legitimacy assessments (cf. Karlsson-Vinkhuyzen and McGee 2013).

Table 7.2 *Legitimacy Assessments of Institutions, structured by actor type.*

	Mean	SE	95% CI	N	Results t-tests
State actors					
CEM	3.026	0.206	[2.590–3.462]	17*	Significantly lower than all other
REN21	3.411	0.259	[2.825–3.998]	10*	
IRENA	3.479	0.149	[3.175–3.783]	29	
IEA	3.678	0.116	[3.440–3.916]	29	Significantly higher than IRENA and CEM
UNFCCC	3.714	0.130	[3.448–3.979]	31	Significantly higher than all but IEA
Nonstate actors					
CEM	2.739	0.149	[2.434–3.045]	29	Significantly lower than all other
IRENA	3.540	0.084	[3.372–3.708]	63	
IEA	3.577	0.083	[3.411–3.743]	72	
REN21	3.655	0.122	[3.407–3.903]	37	
UNFCCC	3.705	0.071	[3.564–3.845]	87	Significantly higher than all but REN21

Source: CLIMENGO Expert Survey 2017–2018.
Notes: * Few respondents within this category were sufficiently familiar with the institution in order to evaluate it on all legitimacy dimensions. Results should be interpreted with this caution in mind. Given the modest sample size, a 90 per cent confidence level is used as the cut-off point for significance testing.

7.6.3 Legitimacy Assessments among Subsets of Stakeholders

We also sought to understand how these legitimacy assessments differ across actors with different backgrounds. As we show in the following, the variation in the data across stakeholder groups qualifies the previous observation: it shows that the formal compliance of an institution with normative legitimacy criteria does not directly translate into stakeholders' legitimacy assessments. By looking for systematic patterns in the legitimacy assessments of different categories of stakeholders, we seek to better understand what shapes such assessments. Further t-tests were therefore conducted in order to explore how different institutions are perceived by different categories of stakeholders and how the issue areas and geographical backgrounds of respondents might affect their assessments of the different institutions.

Table 7.2 shows how state and nonstate actors ranked the different institutions in terms of legitimacy assessments. The means and reported t-tests in Table 7.2 confirm, for both types of actors, the generally observed pattern of a more negative legitimacy assessment of CEM, and a more positive assessment of the UNFCCC, compared to the other institutions. Yet, in addition to this similarity, we also

Table 7.3 *Legitimacy Assessments of Institutions, structured by institution and actor type.*

	Mean	SE	95% CI	N	T-test (t)
CEM					
State actors	3.026	0.206	[2.590–3.462]	17*	
Nonstate actors	2.739	0.149	[2.434–3.045]	29	
					−1.925 (p<0.05)
IEA					
State actors	3.678	0.116	[3.440–3.916]	29	
Nonstate actors	3.577	0.083	[3.411–3.743]	72	
					−1.215 (ns)
IRENA					
State actors	3.479	0.149	[3.175–3.783]	29	
Nonstate actors	3.540	0.084	[3.372–3.708]	63	
					0.727 (ns)
REN21					
State actors	3.411	0.259	[2.825–3.998]	10*	
Nonstate actors	3.655	0.122	[3.407–3.903]	37	
					1.992 (p<0.05)
UNFCCC					
State actors	3.714	0.130	[3.448–3.979]	31	
Nonstate actors	3.705	0.071	[3.564–3.845]	87	
					–0.131 (ns)

Source: CLIMENGO Expert Survey 2017–2018.
Notes: * Few respondents within this category were sufficiently familiar with the institution in order to evaluate it on all legitimacy dimensions. As for CEM and REN21, few state actors are included, and the variances in legitimacy assessments were compared between the largest and smallest group following de Winter (2013). As the variances are relatively equal, the likelihood of Type I error (i.e. observing a false positive result) is low.

observe a slightly different rank-order in legitimacy assessments within both categories of respondents. The IEA (mean 3.678) is ranked significantly higher than IRENA (3.479) amongst state actors, while not being ranked significantly lower than the UNFCCC. Among nonstate actors, REN21 is not assessed as significantly less legitimate than the UNFCCC.

Table 7.3 pairs these figures according to state and nonstate actors' legitimacy assessments for each institution. The means and t-tests further indicate significant differences in the legitimacy assessments among these two actor groups. Assessments of CEM are significantly lower among the surveyed nonstate actors than among the state actors. By contrast, legitimacy assessments of REN21 are significantly more positive among nonstate actors than among state actors. This observation suggests that the inclusion of nonstate actors in an institution plays a role in shaping legitimacy assessments: CEM is the organization with the least access to nonstate actors in our study, while REN21 is the most open one, being a multi-stakeholder network that reaches out to a broad range of nonstate actors in the public and private sectors. Hence, nonstate actors may be more familiar with REN21 so that this institution might be incorporated in their heuristics of what a legitimate climate and energy governance institution could look like. (Tallberg et al. 2014; Lenz and Viola 2017).

Table 7.4 shows how respondents rank institutions differently depending on whether they work in: energy security and technology; climate finance, carbon pricing and mitigation; or adaptation and development. While these categories are partially overlapping, they help distinguish actors according to their main domain (energy, climate, or development).[5] A simple ranking of the mean legitimacy assessment of the five institutions for each category of respondents again shows that CEM is ranked the lowest and the UNFCCC the highest. Yet, no statistically significant differences are detected between the institutions among the adaptation and development respondents. At least for CEM and REN21, this is most likely due to the limited number of respondents. For the climate finance, carbon pricing, and mitigation group, we do observe that the UNFCCC is ranked significantly higher than all other institutions, the IEA significantly higher than IRENA and CEM, and CEM significantly lower than all other institutions. Furthermore, CEM is ranked significantly lower than all other institutions for the energy security and technology respondents. This last finding is counterintuitive, given that CEM has a clear focus on energy and technology questions.

Table 7.5, which rearranges these issue-area-based figures along the five institutions, sheds more light on this observation about CEM. Overall, respondents working with energy security and technology tend to have the most positive legitimacy assessments; those working with adaptation and development have the least positive ones. Moreover, for both CEM and IRENA, the difference between respondents working with energy security and technology and respondents mainly working with other issues is most pronounced. In other words, respondents for whom energy security and technology is most central to their work tend to assess the legitimacy of those institutions that focus most strongly on these issues as particularly more positive than other respondents. For institutions that, next to energy security and technology, also focus on mitigation, climate finance and carbon pricing, development, and adaptation (IEA and UNFCCC), we observe that the

[5] In the survey, respondents were asked to indicate the issues they mainly work with. Most respondents indicated a single issue or multiple issues that fall within one of the three categories we distinguished. The few respondents that combined issues from different categories (e.g. climate finance and energy technology) were included in the samples of both categories.

Table 7.4 *Legitimacy Assessments of Institutions, structured by stakeholders' work focus.*

	Mean	SE	95% CI	N	Results t-tests
Energy security and technology					
CEM	3.044	0.147	[2.743–3.345]	28	Significantly lower than all other institutions
REN21	3.545	0.183	[3.165–3.926]	22*	‡
IRENA	3.632	0.101	[3.428–3.835]	48	
IEA	3.667	0.088	[3.489–3.844]	54	
UNFCCC	3.795	0.085	[3.625–3.964]	56	Significantly higher than all but REN21 (given ‡)
Climate finance, carbon pricing and mitigation					
CEM	2.667	0.171	[2.306–3.027]	18*	Significantly lower than all other institutions
IRENA	3.452	0.093	[3.264–3.640]	43	
REN21	3.545	0.164	[3.202–3.889]	21*	
IEA	3.618	0.102	[3.413–3.824]	46	Significantly higher than IRENA and CEM
UNFCCC	3.761	0.088	[3.585–3.937]	57	Significantly higher than all other institutions
Adaptation and development					(no significant differences between the evaluations of the institutions)
CEM	2.570	0.281	[1.967–3.174]	15*	‡
REN21	3.299	0.200	[2.864–3.735]	13*	No significant differences
IRENA	3.379	0.160	[3.050–3.707]	27	
IEA	3.449	0.149	[3.143–3.755]	26	
UNFCCC	3.511	0.141	[3.227–3.796]	39	

Source: CLIMENGO Expert Survey 2017–2018.
Notes: * Few respondents within this category were sufficiently familiar with the institution in order to evaluate it on all legitimacy dimensions. ‡Following the method of de Winter (2013), no comparison could be made between the mean perceived legitimacy among these respondents. The variance is too high for this small group of observations, compared to variances of the other means in the analysis. Given the modest sample size, a 90 per cent confidence level is used as the cut-off point for significance testing.

legitimacy assessments are not significantly different among respondents working with energy security and technology and those working with climate finance, carbon markets and pricing, and mitigation. This again suggests the importance of the thematic foci of the institutions for the legitimacy assessments of stakeholders. Respondents who work with adaptation and development make the most negative legitimacy assessments for all institutions, even for the UNFCCC, although adaptation and low-carbon development feature prominently on that institution's agenda.

Table 7.5 *Legitimacy Assessments of Institutions, structured by institution and stakeholders' work focus.*

	Mean	SE	95% CI	N	T-test
CEM					
1. Energy security and technology	3.044	0.147	[2.743–3.345]	28	1 vs. 2: t=2.569, p<0.01 1 vs. 3: ‡
2. Climate finance, carbon pricing and mitigation	2.667	0.171	[2.306–3.027]	18*	2 vs. 1: t=-2.208, p<0.05 2 vs. 3: ‡
3. Adaptation and development	2.570	0.281	[1.967–3.174]	15*	3 vs. 1: ‡ 3 vs. 2: ‡
IEA					
1. Energy security and technology	3.667	0.088	[3.489–3.844]	54	1 vs. 2: t=0.551, ns 1 vs. 3: t=2.462, p<0.01
2. Climate finance, carbon pricing and mitigation	3.618	0.102	[3.413–3.824]	46	2 vs. 1: t=-0.477, ns 2 vs. 3: t=1.661, p<0.1
3. Adaptation and development	3.449	0.149	[3.143–3.755]	26	3 vs. 1: t=-1.469, p<0.1 3 vs. 2: t=-1.140, ns
IRENA					
1. Energy security and technology	3.632	0.101	[3.428–3.835]	48	1 vs. 2: t=1.779, p<0.05 1 vs. 3: t=2.500, p<0.001
2. Climate finance, carbon pricing and mitigation	3.452	0.093	[3.264–3.640]	43	2 vs. 1: t=-1.932, p<0.05 2 vs. 3: t=0.787, ns
3. Adaptation and development	3.379	0.160	[3.050–3.707]	27	3 vs. 1: t=-1.586, p<0.1 3 vs. 2: t=-0.459, ns
REN21					
1. Energy security and technology	3.545	0.183	[3.165–3.926]	22*	1 vs. 2: t=0.003, ns 1 vs. 3: t=2.569, p<0.01
2. Climate finance, carbon pricing and mitigation	3.545	0.164	[3.202–3.889]	21*	2 vs. 1: t=-0.002, ns 2 vs. 3: t=1.496, p<0.1
3. Adaptation and development	3.299	0.200	[2.864–3.735]	13*	3 vs. 1: t=-1.230, ns 3 vs. 2: t=-1.230, ns
UNFCCC					
1. Energy security and technology	3.795	0.085	[3.625–3.964]	56	1 vs. 2: t=0.398, ns 1 vs. 3: t=3.355, p<0.001
2. Climate finance, carbon pricing and mitigation	3.761	0.088	[3.585–3.937]	57	2 vs. 1: t=-0.388, ns 2 vs. 3: t=2.849, p<0.01

Table 7.5 (*cont.*)

	Mean	SE	95% CI	N	T-test
3. Adaptation and development	3.511	0.141	[3.227–3.796]	39	3 vs. 1: t=-2.018, $p<0.05$ 3 vs. 2: t=-1.777, $p<0.05$

Source: CLIMENGO Expert Survey 2017–2018.
Notes: * Few respondents within this category were sufficiently familiar with the institution in order to evaluate it on all legitimacy dimensions. ‡ Given the low N of one of the categories in the comparison, the variances in legitimacy assessments were compared between the largest and smallest group following de Winter (2013). When the variances are relatively equal or when the variance is smaller in the category with the lowest N, the likelihood of Type I error (i.e. observing a false positive result) is low. In these cases, the result of the t-test is presented. Where this criterion is not met, the result of the t-test is omitted.

These observations are in line with one of our aforementioned expectations, namely that differences in legitimacy assessments could stem from differences in norms and values amongst communities of stakeholders working on similar issues, or from differences in the institutions that they are familiar with and use as heuristics or prototypes. According to the cognitive model, the observed positive assessments of respondents working on issues of energy security and technology would be explained by their higher level of familiarity with institutions that fulfil fewer criteria for normative legitimacy, which in turn would reflect in the heuristics they use to compare the five institutions in our study against. In comparison to those, they assess the five climate-energy institutions in our study more favourably. For example, if respondents working mainly with energy issues take institutions such as OPEC and the International Energy Forum as reference points, the five institutions in this study could be considered more legitimate as they fulfil more normative criteria, particularly in terms of openness and transparency.

By contrast, those respondents working on issues of adaptation and development tend to be familiar with institutions that fulfil rather more criteria for normative legitimacy. Compared to such prototypes, they would assess the five climate-energy institutions less favourably. For example, the Global Environment Facility and the Green Climate Fund that focus more on development issues may constitute such reference institutions (in terms of heuristics) for the respondents who work mainly on adaptation and development. Particularly the Global Environment Facility has been discussed as a potential role model for other international institutions due to its inclusiveness and openness toward a diversity of actors (Streck 2001).

In summary, we can expect the prototype institutions to differ considerably across groups of respondents, which could explain a large part of the difference in legitimacy assessments that we found for our sample. This said, this connection needs to be further corroborated, since our survey data does not include information on the heuristic that respondents had in mind when assessing the five institutions. The focus of our study, thus, remains exploratory and descriptive, yet it suggests avenues for further explanatory research.

Finally, we explored whether significant variations can be observed in the legitimacy assessments by respondents from countries with different economic backgrounds, or from different geographical regions (Africa, Asia and Pacific, Latin America and Caribbean, and European and other Western countries). For the UNFCCC, we indeed observe such a significant difference. Respondents from high-income countries (as categorized by the World Bank) perceive the UNFCCC as significantly more legitimate (mean= 3.773) than respondents from other countries (mean= 3.563; t= 2.912; p=0.002). Moreover, no significantly different views about the legitimacy of the UNFCCC are observed for respondents from middle- and low-income countries. Neither did we observe a significant difference in evaluations of the other institutions when we grouped respondents by national-income category.[6]

This is rather surprising, given that both norms and values and what institutions respondents are familiar with would be expected to vary with the geographic origin of the respondent. It may be that many of the respondents are international elites and have therefore been socialized or self-selected into similar norms, and are hence used to similar international institutions. Thus, we might indeed be looking at dynamics of a transnational elite that is divided by professional focus, rather than by nationality – since we did observe distinctions in legitimacy assessments across respondents from different sectors and types of organizations. In fact, previous research supports this assumption: Verhaegen et al. 2018, for instance, showed that there is more variation in legitimacy perceptions of global governance institutions between elites of different societal sectors than between elites from different countries.

7.7 Conclusions

The aim of this chapter was to provide a first mapping of stakeholders' assessments of the legitimacy of five key institutions governing the climate-energy nexus. Against the backdrop of considerable institutional complexity, and scarce resources amongst public and private actors to enhance participation in global

[6] Additionally, we observed that a one-way ANOVA does not show significant differences between respondents from different world regions (based on continents) in terms of legitimacy assessments of any of the institutions. There were too few respondents answering all questions about CEM and REN21 to allow for a comparison between member states and non-member states.

governance, we wanted to better understand to what extent key institutions in global climate and energy governance are seen as legitimate by key stakeholders.

The analyses showed that, on the one hand, there are many similarities in the legitimacy assessments of the five institutions we put under scrutiny – with the mean legitimacy assessments ranging between 2.845 (CEM) and 3.707 (UNFCCC) on a scale from 1 to 5. On the other hand, we also found systematic differences across stakeholders of different types, working with different issues and – to a limited extent – coming from different countries.

Specifically, we observed that CEM is systematically assessed as the least legitimate, and the UNFCCC as the most legitimate, of the five institutions. Second, our analyses showed that the legitimacy assessments of nonstate actors are more positive toward institutions that are more inclusive toward this type of stakeholders. Third, we observed that stakeholders working with energy security and technology, and those working with climate finance, carbon pricing, and mitigation have more positive legitimacy assessments of institutions that more strongly focus on their issues. By contrast, respondents working with adaptation and development issues assessed the legitimacy of the selected institutions more negatively than the other respondents, even for the UNFCCC, which is the global institution in our sample that most strongly engages with these issues. We can only speculate about the reasons for this. Our study has highlighted the possibility that differences in these communities' norms, values, and experiences contribute to different heuristics being used to make assessments. Yet, whether such differences ultimately stem from processes of socialization or whether they are rather due to functionalist or rationalistic reasons is a pertinent question for future research.

These limitations notwithstanding, the results of our unique survey allow us to draw a set of novel conclusions. First, the results appear to support the view that stakeholders do not adequately disentangle their legitimacy assessments of individual institutions that have similar functions and overlapping mandates. Perhaps this is a reflection of the relatively high level of coordination in the renewable energy subfield (Sanderink, Chapter 4), which means that institutions in this subfield interact extensively with one another and thereby make it difficult for stakeholders to distinguish their respective performance. Within each category of stakeholders, we found comparable assessments and similar legitimacy rankings for these institutions, albeit with some small significant differences. We have reasons to believe that, in order to navigate in a very complex governance field, the surveyed stakeholders form their assessments based on a comparison with institutions that are familiar to, or valued by them. Faced with incomplete information and due to bounded rationality, stakeholders use mental shortcuts to make such comparisons and base their legitimacy assessments thereupon.

Second, the differences in legitimacy assessments found between governmental versus nonstate actors, and across stakeholders working on different issue areas, suggest that international institutions have to pursue different legitimation strategies for different audiences (Gronau and Schmidtke 2016; Bäckstrand and Söderbaum 2018; Verhaegen et al. 2018). Knowing one's audiences is particularly important for institutions that seek to establish and maintain legitimacy in an increasingly crowded field.

Third, and more generally, a stakeholder's level of familiarity with an institution appears to be linked to a more positive assessment of legitimacy. It therefore does not come as a surprise that international institutions increasingly engage in outreach activities, especially on social media – in order to promote their work to a diversity of actors and seek input from them.

While these findings advance the research frontier on legitimacy, the reliance on survey data comes with the usual set of shortcomings, which means that the findings need to be confirmed in future studies. First, there is always the possibility that respondents either think of the institution as a whole or just the secretariat or another institutional body, which makes a straight-off comparison difficult (Zaum 2013). Second, we followed a notion of cognitive legitimacy whereby respondents' assessments are based on comparisons with heuristic or prototype institutions. Which particular heuristic institutions that are used by respondents lies beyond the scope of this study. The first explorative results offered in this chapter should therefore be examined in further, especially interview-based, studies. These could also delve deeper into questions such as how expertise, which is considered key for international institutions in the climate-energy nexus, is conceptualized by stakeholders. Finally, the links to other explanatory variables, such as resources, staffing, or relations to other institutions, should be pursued to further understand assessments of legitimacy.

This study also opens up empirical avenues for further research. It provided a first mapping of stakeholders' perceptions of nine legitimacy dimensions across five institutions for one particular subfield. An examination of institutions from other subfields could provide insights into how the level of coherence within institutional complexes affects issues of legitimacy. Next steps could also measure differences in how stakeholders view the relative importance of the nine dimensions, or other dimensions of legitimacy not included in this study, to also learn about the sociological legitimacy of the institutions. An interesting and policy-relevant line of inquiry is how low assessments of certain dimensions of legitimacy can be, and amongst which groups of stakeholders, before the institution faces a legitimacy crisis. An answer to that question would, however, require a much larger survey of stakeholders. One limitation of this study has been that the survey includes too few cases (N) in order to do a multivariate analysis that allows comparing the relationship between actor type, issue area, and geographical origin on the one hand, and assessments of legitimacy of the five global climate and energy institutions on the other. A larger research effort would be needed to address this limitation. Such a research effort would also be useful to provide a more fine-grained analysis of differences in legitimacy assessments between different categories of nonstate actors, such as businesses and civil society organizations.

Finally, considerations of legitimacy will always be of major importance for policy makers when deciding on which institution to work with and invest in. Institutional complexity affects these conditions, as institutions and their legitimacy have become highly entangled. Therefore, further research questions, such as about the role of legitimation and delegitimation strategies under institutional complexity, merit further enquiry, as such strategies are likely to affect institutions differently, depending on the norms, values, and experiences of the legitimacy-granting communities.

7.8 References

Agné, H. 2018. Legitimacy in Global Governance Research: How Normative or Sociological Should It Be? In *Legitimacy in Global Governance: Sources, Processes, and Consequences*, edited by J. Tallberg, K. Bäckstrand, and J. A. Scholte. Oxford: Oxford University Press.

Alter, K. and Meunier, S. 2009. The Politics of International Regime Complexity. *Perspectives on Politics* 7(1), 13–24.

Andresen, S. and Hey, E. 2005. The Effectiveness and Legitimacy of International Environmental Institutions. *International Environmental Agreements: Politics, Law and Economics* 5(3), 211–226.

Armingeon, K and Ceka, B. 2014. The Loss of Trust in the European Union during the Great Recession since 2007: The Role of Heuristics from the National Political System. *European Union Politics* 15(1), 82–107.

Bäckstrand, K. and Söderbaum, F. 2018. Legitimation and Delegitimation in Global Governance: Discursive, Institutional, and Behavioural Practices. In *Legitimacy in Global Governance: Sources, Processes, and Consequences*, edited by J. Tallberg, K. Bäckstrand and J. A. Scholte. Oxford: Oxford University Press.

Bäckstrand, K., Zelli, F., and Schleifer, P. 2018. The Legitimacy and Accountability in Polycentric Climate Governance. In *Governing Climate Change: Polycentricity in Action*, edited by A. Jordan, D. Huitema, H. van Asselt, and J. Forster. Cambridge: Cambridge University Press.

Beetham, D. 1991. *The Legitimation of Power*. Basingstoke: Palgrave.

Bernstein, S. 2005. Legitimacy in Global Environmental Governance. *Journal of International Law and International Relations* 1(1–2), 139–166.

Biermann, F., Pattberg P., van Asselt, H. and Zelli, F. 2009. The Fragmentation of Global Governance Architectures: A Framework for Analysis. *Global Environmental Politics* 9(4), 14–40.

Bodansky, D. 1999. The Legitimacy of International Governance: A Coming Challenge for International Environmental Law? *American Journal of International Law* 93(3), 596–624.

Brun, A. 2016. Conference Diplomacy: The Making of the Paris Agreement. *Politics and Governance* 4(3), 115–123.

CEM (Clean Energy Ministerial). 2018. About the Clean Energy Ministerial. Available at www.cleanenergyministerial.org/about-clean-energy-ministerial.

Dellmuth, L. M. and Tallberg, J. 2015. The Social Legitimacy of International Organizations: Interest Representation, Institutional Performance, and Confidence Extrapolation in the United Nations, *Review of International Studies* 41(3), 451–475.

Fabrigar, L. R. and Wegener, D. T. 2012. *Exploratory Factor Analysis*. New York: Oxford University Press.

Flockhart, T. 2006. 'Complex Socialization': A Framework for the Study of State Socialization, *European Journal of International Relations* 12(1), 89–118.

Gibson, J. L. and Caldeira, G. A. 1998. Changes in the Legitimacy of the European Court of Justice: A Post-Maastricht Analysis. *British Journal of Political Science* 28(1), 63–91.

Greenhill, B. 2010. The Company You Keep: International Socialization and the Diffusion of Human Rights Norms, *International Studies Quarterly* 54(1), 127–145.

Gronau, J. and Schmidtke, H. 2016. The Quest for Legitimacy in World Politics – International Institutions' Legitimation Strategies. *Review of International Studies* 42(3), 535–557.

Gutner, T. and Thompson, A. 2010. The Politics of IO Performance: A Framework *The Review of International Organizations* 5(3), 227–248.

IEA (International Energy Agency). 2018a. Countries. Available at www.iea.org/countries/.

IEA (International Energy Agency). 2018b. About. Available at www.iea.org/about/.

IEA (International Energy Agency). 2018c. Structure. Available at www.iea.org/about/structure/.

IRENA (International Renewable Energy Agency). 2020a. Irena Membership. Available at irena.org/irenamembership.

IRENA (International Renewable Energy Agency). 2020b. Vision and Mission. Available at irena.org/statutevisionmission.

IRENA (International Renewable Energy Agency). 2020c. IRENA Coalition of Action. Available at coalition.irena.org/

Karlsson-Vinkhuyzen, S. I. and McGee, J. 2013. Legitimacy in an Era of Fragmentation: The Case of Global Climate Governance. *Global Environmental Politics* 13(3), 56–78.

Karlsson-Vinkhuyzen, S. I. and Vihma, A. 2009. Comparing the Legitimacy and Effectiveness of Global Hard and Soft Law: An Analytical Framework. *Regulation & Governance* 3(4), 400–420.

Keohane, R. O. 2011. Global Governance and Legitimacy. *Review of International Political Economy* 18(1), 99–109.

Lenz, T. and Viola, L. A. 2017. Legitimacy and Institutional Change in International Organisations: A Cognitive Approach. *Review of International Studies* 43(5), 939–961.

Mena, S. and Palazzo, G. 2012. Input and Output Legitimacy of Multi-Stakeholder Initiatives. *Business Ethics Quarterly* 22(3), 527–556.

Nasiritousi, N., Hjerpe, M., and Bäckstrand, K. 2016. Normative Arguments for Non-State Actor Participation in International Policymaking Processes: Functionalism, Neocorporatism or Democratic Pluralism? *European Journal of International Relations* 22(4), 920–943.

Nasiritousi, N. and Linnér, B-O. 2016. Open or Closed Meetings? Explaining Nonstate Actor Involvement in the International Climate Change Negotiations. *International Environmental Agreements: Politics, Law and Economics* 16(1), 127–144.

Raines, S. S. 2003. Perceptions of Legitimacy and Efficacy in International Environmental Management, *Global Environmental Politics* 3(3), 47–73.

REN21 (Renewable Energy Policy Network for the 21st Century). 2018a. About us. Available at www.ren21.net/about-ren21/about-us/.

REN21 (Renewable Energy Policy Network for the 21st Century). 2020. Members. Available at www.ren21.net/your-network/the-ren21-members/.

REN21 (Renewable Energy Policy Network for the 21st Century). 2018b. REN21 Highlights. Available at www.ren21.net/about-ren21/ren21-highlights/.

Scharpf, F. 1999. *Governing in Europe: Effective and Democratic?* Oxford: Oxford University Press.

Scholte, J. A. and Tallberg, J. 2018. Theorizing the Institutional Sources of Global Governance Legitimacy. In *Legitimacy in Global Governance: Sources, Processes, and Consequences*, edited by J. Tallberg, K. Bäckstrand, and J. Scholte. Oxford: Oxford University Press.

Streck, C. 2001. The Global Environment Facility – a Role Model for International Governance? *Global Environmental Politics* 1(2), 71–94.

Symons, J. 2011. The Legitimation of International Organisations: Examining the Identity of the Communities that Grant Legitimacy. *Review of International Studies* 37(5), 2557–2583.

Szulecki, K., Pattberg, P., and Biermann, F. 2011. Explaining Variation in the Effectiveness of Transnational Energy Partnerships. *Governance* 24(4), 713–736.

Tallberg, J., Sommerer, T., Squatrito, T., et al. 2014. Explaining the Transnational Design of International Organizations. *International Organization* 68(4), 741–774.

Underdal, A. 2002. One Question, Two Answers. In *Environmental Regime Effectiveness: Confronting Theory with Evidence*, edited by E. L. Miles, A. Underdal, S. Andresen, J. Wettestad, J. B. Skjærseth, and E. M. Carlin. Cambridge, MA: MIT Press.

UNFCCC (United Nations Framework Convention on Climate Change). 2018. UNFCCC Process. Available at https://unfccc.int/process.

Van de Graaf, T. 2013. Fragmentation in Global Energy Governance: Explaining the Creation of IRENA. *Global Environmental Politics* 13(3), 14–33.

Verhaegen, S., Scholte, J. A., and Tallberg, J. 2018. Elite Perceptions of Legitimacy in Global Governance. Paper presented at the Annual Convention of the International Studies Association, 4–7 April, San Francisco.

Zaum, D. 2013. International Organizations, Legitimacy, and Legitimation. In *Legitimating International Organizations*, edited by D. Zaum. Oxford: Oxford University Press.

8

The Performance of the Climate-Energy Nexus

Assessing the Effectiveness of the Institutional Complexes on Renewable Energy, Fossil Fuel Subsidy Reform, and Carbon Pricing

NAGHMEH NASIRITOUSI, LISA SANDERINK, JAKOB SKOVGAARD, HARRO VAN ASSELT, CLEO VERKUIJL, AND OSCAR WIDERBERG

8.1 Introduction

What does institutional complexity mean for performance in the climate-energy nexus? As previous chapters have shown, the nexus is made up of a diverse set of institutions that have overlapping mandates and functions. Chapter 3 showed how the institutional complex varies at the meso level, and Chapters 4–6 explored the interactions between different institutions in three selected subfields: renewable energy, fossil fuel subsidy reform, and carbon pricing. Given the large number of institutions that operate and interact in these fields, several questions arise about their performance and environmental effectiveness: what are the consequences of this intricate web of institutions for the performance of the institutional complex of the climate-energy nexus? Is institutional complexity a requirement for effective problem-solving, or does it hamper effectiveness? What management options exist for making the institutional complex at the climate-energy nexus more effective? Considering the magnitude of the climate- and energy-related challenges, the answers to these questions are of great importance to both scholars and practitioners.

Based on these questions, and building on the previous chapters, the aim of this chapter is to assess the effectiveness of each of the three subfields as well as to discuss the overall performance of the institutional complex of the climate-energy nexus. As outlined in Chapter 2 and elaborated on in the next section, effectiveness here refers to how well institutions perform in terms of achieving goals that they have been tasked to fulfil. By examining the outputs, outcomes, and impacts of the three subfields, the chapter shows both the advantages and the disadvantages of institutional complexity of the climate-energy nexus for achieving effectiveness. It further shows that, despite the difficulties with evaluating effectiveness under institutional complexity, such an assessment is a worthwhile exercise in order to identify management options – i.e. options for formally regulating the linkage between institutions – for the climate-energy nexus.

The chapter proceeds as follows. The next section discusses the concept of effectiveness and the challenges to analyzing the effectiveness of institutions, especially when they have overlapping mandates and are interlinked. Thereafter, our methodology section outlines how, in order to respond to these challenges, our research relies on a two-track approach, integrating assessments by researchers and interviews with key stakeholders. Based on this information, we evaluate the outputs, outcomes, and impacts of institutions within each subfield. Thereafter, we examine what the consequences of institutional complexity are for the subfield in question. The insights gained from this analysis are then used to outline management options for the institutions of the climate-energy nexus. The final section concludes with discussing implications of our findings for the governance of the nexus at large.

8.2 Conceptualizing Effectiveness

As discussed in Chapter 2, effectiveness can be evaluated in different ways. Easton (1965) suggested measuring effectiveness across three dimensions: output, outcome, and impact. *Output* concerns performance in terms of what the institution produces, for example issuing regulations (binding or non-binding), producing reports, conducting research, organizing meetings, providing funding, offering training, etc. (Szulecki et al. 2011). *Outcome* relates to whether the institution produces behavioural changes, for example in terms of whether it increases the level of cooperation and compliance amongst members, for instance by improving learning and modifying incentives (Underdal 2002; Gutner and Thompson 2010). To determine an institution's *impact* implies assessing the extent to which the institution contributes to alleviating the problem it was tasked to resolve (Underdal 2002). Impacts may include effects that are positive or negative, direct or indirect, intentional or unintentional, and these can be short-, medium-, or long-term (Alcamo 2017). This threefold understanding implies that effectiveness is a stronger term than performance, since institutions can perform well in terms of output but nevertheless not achieve the intended impacts necessary for goal attainment.

Measuring effectiveness becomes increasingly difficult as the number of institutions rises. Even just for one institution, assessing effectiveness is challenging because of the need to establish causality between the output of the institution, the behavioural change among the target actors, and the impact on the problem that the institution was tasked to solve. This challenge is multiplied under institutional complexity because of the question of attribution, namely which institutions are responsible for observed impacts in a web of institutions with overlapping mandates? In short, under institutional complexity, the difficulties involved in assessing effectiveness are compounded by challenges in identifying the division of labour between institutions (Alter and Meunier 2009).

Moreover, when evaluating impact for a field with multiple institutions, the analysis shifts from assessing goal attainment for individual institutions to assessing how the work of multiple institutions affects an overall goal, such as the fulfilment of Sustainable Development Goal (SDG) 7 on sustainable energy in the case of the renewable energy subfield. This approach is different from what can be found in much of the previous literature on effectiveness, where the focus is either on assessing institutional or environmental effectiveness (Underdal 2002; Gutner and Thompson 2010; Tallberg et al. 2016). Analyses of institutional effectiveness look at institutional performance, also including assessments of output legitimacy, such as the one presented in Chapter 7. Studies on environmental effectiveness, on the other hand, look at the extent to which specific institutions have an impact on environmental indicators. In contrast, the analysis in this chapter looks at the extent to which the collective contributions of individual institutions within a subfield are successful in fulfilling common goals in the subfield.

Some studies seek to circumvent the challenge of identifying outcomes and impacts of institutions by simply focusing on outputs (Szulecki et al. 2011; Tallberg et al. 2016). By examining outputs, these scholars assess the performance of institutions and thereby look at *potential* effectiveness. Alternatively, some studies analyze effectiveness by examining whether institutions are producing the outputs that could be expected, given the functions that they have (Pattberg et al. 2012; Chan et al. 2018). However, these approaches are at best useful as first steps and approximations for assessing the actual effects of institutions on the governance of particular issue areas.

The approach employed in this chapter instead seeks to link outputs to observed outcomes and impacts. For each subfield, our approach identifies specific outputs and discusses possible outcomes and impacts. While imperfect due to knowledge limitations, this approach makes explicit how assessments of effectiveness are made and thereby allows for critical reflection and learning about the cause and effect of institutional consequences. The aim of the analysis is hence not to show whether the institutions are effective or not but to discuss and specify in what ways they could be seen as effective (or not) and how institutional complexity affects effectiveness. The value of this approach lies in its context-specific analysis of outputs, outcomes, and impacts for each subfield and in deriving suitable management options to enhance effectiveness. Our own assessments are complemented by interview data from a range of stakeholders with high familiarity of institutions working within these subfields, as explained in the next section.

Assessing effectiveness across the three subfields (renewable energy, fossil fuel subsidy reform, and carbon pricing) was carried out by analyzing key documents and reports, as well as conducting semi-structured interviews with various experts. For the document analysis, we focused on academic journal articles but also included grey literature such as reports from international organizations and non-governmental organizations. For the interviews, we approached, for each subfield, representatives from national governments, international organizations, NGOs and academia. The interviewees were selected based on thematic expertise and knowledge of the institutions in each case study, with a view to cover a wide variety of actor types, countries and sectors. In total, thirty-eight interviews were carried out across the three subfields.

The interviews covered the following aspects: (1) the overall degree of effectiveness in the subfield; (2) possible bottlenecks that may hamper effectiveness; (3) influence of institutional complexity on effectiveness; and (4) management options by particular institutions to improve the effectiveness of the subfield. The analyses are based on a careful assessment of the data retrieved from the document analyses and the interviews. More specifically, output effectiveness was mainly determined based on the analyses in Chapters 4–6, whereas estimations of outcome and impact were mostly derived from academic and grey literatures. Findings about the influence of institutional complexity on effectiveness as well as management options are mainly based on the experts' views.

In what follows, we examine effectiveness, the consequences of institutional complexity and management options for each subfield at the meso level. The concluding section offers a comparison of the three subfields and draws out the implications of our findings for the performance of the institutional complex of the climate-energy nexus.

8.3 Assessments of Effectiveness for the Renewable Energy Subfield

8.3.1 Assessment of Outputs, Outcomes, and Impacts

A sustainable energy future hinges on a worldwide uptake of renewable energy. Many of the institutions that operate in the renewable energy subfield relate their work both to the SDG 7 target on clean energy and to the Paris Agreement's temperature target (Chapter 4). This section assesses the effectiveness of the densely crowded subfield for renewable energy.

First, with regard to output, Chapter 4 showed that the majority of renewable energy institutions focus on information-sharing through dissemination of research and publishing reports. Consequently, the renewable energy subfield displays a diversity of knowledge and expertise on energy sources and technologies from a wide range of perspectives, which is frequently shared at various meetings, conferences, and platforms.[1] In addition, there are a fair number of renewable energy institutions working toward capacity-building and project implementation (see also Chapter 4). These particularly focus on deploying renewables for the purpose of expanding energy access in the developing world. Financing schemes, regulations, standards, and commitments are produced to a lesser extent.

Second, in terms of outcome, awareness and capacity are growing along a wide spectrum of stakeholders. For example, via renewable energy institutions, national governments are increasingly sharing experiences and taking note of best practices.[2] Simultaneously, nonstate engagement is spreading, which is illustrated by the growing number of private initiatives and multi-stakeholder partnerships for renewable energy (see Chapter 4). Businesses, trade associations, financial institutions, NGOs, and other civil society organizations show increasing interest in renewables. In short, renewable energy institutions appear to stimulate if not behavioural, then attitudinal changes amongst their members and beyond.

Third, assessing the level of goal attainment or problem-solving capacity suggests a low degree of effectiveness. Despite 2017 being a record-breaking year for the share of renewables in the global energy mix, the growth rate is currently falling short of meeting either the ‘substantial increase’ by 2030, as targeted by SDG 7, or the 2-degrees target set by the Paris Agreement (IRENA 2018; REN21 2018; United Nations 2018). It is difficult to determine the level of effectiveness based on broad perceptions, let alone for an institutional complex that includes such a high number of different institutions. This notwithstanding, the currently inadequate growth rate for renewables suggests that the subfield’s institutional complex has suboptimal performance.[3]

Why is this the case? We could identify various bottlenecks through our interviews and literature review. First, the renewable energy subfield inherited the dominance of national policy making in global energy governance (Karlsson-Vinkhuyzen et al. 2012; Röhrkasten 2015; Van de Graaf and Zelli 2016). Even though renewable energy may be a less strategic issue for national security compared to traditional sources of energy, national governments have remained

[1] Interviews with T. Van de Graaf, Professor, Ghent Institute for International Studies, Ghent University, 13 July 2018; and G. Fernandez Ludlow, Director for Climate Change, E. M. del Pilar Casamadrid Gutiérrez, Director for the Environment and J. Alarcón González, Head of Department for Climate Change, Ministry of Foreign Affairs, Mexico, 3 July 2018.

[2] Interviews with G. Fernandez Ludlow, Director for Climate Change, E. M. del Pilar Casamadrid Gutiérrez, Director for the Environment and J. Alarcón González, Head of Department for Climate Change, Ministry of Foreign Affairs, Mexico, 3 July 2018; and M. Raamat, Counsellor on International Relations and Energy, Ministry of Environment, Estonia, 21 September 2018.

[3] Interviews with B. Sovacool, Professor of Energy Policy, Science Policy Research Unit, University of Sussex, 10 May 2018; and B. Hoskuldsson, Lead Partnership Specialist, Sustainable Energy for All (SEforALL), 21 September 2018.

hesitant to give up sovereignty.[4] Second, the subfield is steered by three different challenges, each of which controversial in its own right – energy security, energy access, and environmental sustainability – resulting in trade-offs and potential conflicts across institutions (see Chapter 4; Newell et al. 2011; and Röhrkasten 2015). Furthermore, there is no clear definition of what constitutes a renewable source of energy, leading to further controversies, for example related to nuclear power, bioenergy, and hydropower.[5] These aspects at least hinder the effectiveness of the subfield as a whole in terms of accelerating a renewables uptake. Several other bottlenecks relate to institutional complexity and are therefore discussed in the next subsection.

8.3.2 Consequences of Institutional Complexity: What Are the Implications for Renewable Energy?

To what extent can the low degree of effectiveness be attributed to the institutional complexity of the renewable energy subfield? Compared to the other two cases in this edited volume, carbon pricing and fossil fuel subsidy reform, the renewable energy subfield can be regarded as highly institutionally complex. The subfield is densely populated by a diverse set of institutions, which do not only differ in terms of their structural characteristics but also with respect to the functions they perform, the sources of energy and technologies they cover, and the challenges they seek to address (see Chapter 4). While this complexity makes it difficult to establish a causal relationship with the level of effectiveness, our interviewees on balance expect more advantages than disadvantages with institutional complexity.[6]

On the one hand, institutional complexity is considered to support effectiveness in two ways. First, the variety of institutions involved renders more comprehensive information available from a wide range of perspectives on renewable energy sources and technologies as well as on developments and innovations in the field.[7]

[4] Interview with T. Van de Graaf, Professor, Ghent Institute for International Studies, Ghent University, 13 July 2018.

[5] Interviews with S. Gsänger, Secretary-General, World Wind Energy Association (WWEA), Vice Chair, Renewable Energy Policy Network for the 21st Century (REN21), 9 May 2018; S. Singer, Advisor Global Energy Policies, Climate Action Network International, 10 May 2018; and S. Röhrkasten, Scientific Project Lead Pathways to Sustainable Energy, Institute for Advanced Sustainability Studies (IASS), 17 May 2018; and T. Van de Graaf, Professor at Ghent Institute for International Studies, Ghent University, 13 July 2018.

[6] Interviews with F. Van der Vleuten, Senior Energy Expert, Climate Team Ministry of Foreign Affairs, the Netherlands, 5 June 2018; B. Hoskuldsson, Lead Partnership Specialist, Sustainable Energy for All (SEforALL), 21 September 2018; and L. Williamson, Outreach and Communication Manager, Renewable Energy Policy Network for the 21st Century (REN21), 27 September 2018.

[7] Interviews with B. Sovacool, Professor of Energy Policy, Science Policy Research Unit, University of Sussex, 10 May 2018; G. Fernandez Ludlow, Director for Climate Change, E. M. del Pilar Casamadrid Gutiérrez, Director for the Environment and J. Alarcón González, Head of Department for Climate Change, Ministry of Foreign Affairs, Mexico, 3 August 2018; and M. Raamat, Counsellor on International Relations and Energy, Ministry of Environment, Estonia, 18 September 2018.

Second, institutional complexity provides the opportunity to disaggregate an intricate issue such as renewable energy into smaller challenges and to work on them in parallel with different degrees of progress.[8] Such a compartmentalizing approach has proven to be effective for the climate change realm and may be particularly suitable for a subfield such as renewable energy, characterized by the diversity of energy sources and technologies and differing challenges to tackle.

On the other hand, institutional complexity may turn out problematic for effectiveness in several ways. First, the interviewees express concerns about duplication of work and conflictive impacts among the renewable energy institutions.[9] With several institutions working on similar issues, it is sometimes unclear whether there are overlaps, or worse, incongruences, trade-offs, and conflicts between institutions (Biermann et al. 2009). As a consequence, it is difficult for national governments to decide which organizations, partnerships, and initiatives to participate in, and for the institutions themselves to identify thematic and functional gaps that need to be filled. Second, there exists competition over resources, visibility, sphere of influence, and media attention. This competition may involve institutions that target different renewable energy sources, but also institutions from related issue areas such as energy efficiency.[10] Third, there is no single institution with universal membership in the renewable energy subfield.[11] Such an institutional umbrella may, according to some observers, be ultimately necessary to achieve the common goal to substantially increase the share of renewables in the global energy mix. However, with 160 states as members and 23 in accession by 2019, IRENA is well on its way to positioning itself as one and to continue its unique multilateral success story in global (renewable) energy governance (Röhrkasten and Westphal 2013; Urpelainen and Van de Graaf 2015).

Yet, there is also a different perspective. Various scholars have recently argued that the emerging global transition toward renewable energy is not the result of deliberate and integrated international cooperation, but rather the result of an

[8] Interviews with T. Van de Graaf, Professor, Ghent Institute for International Studies, Ghent University, 13 July 2018; and L. Williamson, Outreach and Communication Manager, Renewable Energy Policy Network for the 21st Century (REN21), 27 September 2018.

[9] Interviews with F. Van der Vleuten, Senior Energy Expert, Climate Team Ministry of Foreign Affairs, The Netherlands, 5 June 2018; G. Fernandez Ludlow, Director for Climate Change, E. M. del Pilar Casamadrid Gutiérrez, Director for the Environment and J. Alarcón González, Head of Department for Climate Change, Ministry of Foreign Affairs, Mexico, 3 August 2018; and B. Hoskuldsson, Lead Partnership Specialist, Sustainable Energy for All (SEforALL), 21 September, 2018.

[10] Interviews with B. Sovacool, Professor of Energy Policy, Science Policy Research Unit, University of Sussex, 10 May 2018; S. Singer, Senior Advisor Global Energy Policies, Climate Action Network International, 10 May 2018; F. Van der Vleuten, Senior Energy Expert, Climate Team Ministry of Foreign Affairs, The Netherlands, 5 June 2018; B. Hoskuldsson, Lead Partnership Specialist, Sustainable Energy for All (SEforALL), 21 September 2018.

[11] Interview with T. Van de Graaf, Professor, Ghent Institute for International Studies, Ghent University, 13 July 2018.

organic proliferation of bottom-up initiatives (e.g. Aklin and Urpelainen 2018; Meckling 2019). Although it is difficult to assess the overall consequences of institutional complexity on effectiveness in the renewable energy subfield, our findings tend to support the argument that the current institutionally complex structure seems fitting, and perhaps even required, for the renewable energy subfield (see also Young 2002). Furthermore, there may be less of a need for institutional integration today than prior to 2015: with the Paris Agreement and SDG 7 agreed upon, *'discursively there is in any case a high degree of consensus'*.[12]

8.3.3 Management Options for the Renewable Energy Subfield

The general view among climate and energy experts interviewed is that the renewable energy subfield is functioning fairly well, institution-wise, partly guided by the targets and principles presented in the Paris Agreement and SDG 7.[13] This notwithstanding, there is a need for increased coordination among the renewable energy institutions to resolve the potentially negative implications of institutional complexity and, ultimately, to achieve targets to substantially accelerate the worldwide uptake of renewables.

First, it is necessary to map out existing renewable energy institutions, and their functions and targeted impacts, and to keep track of the progress being made.[14] This will help to prevent and resolve duplication of work and conflictive impacts and to identify docking points and gaps that need to be addressed among existing institutions. To clarify the latter, there is no need for new institutions trying to reinvent the wheel, but rather to find ways for collaboration to strengthen the overall outcome.[15] Second, with a plethora of knowledge and expertise comes a variety of scenarios, statistics, and data, based on a range of different methodologies and definitions that are not always compatible across institutions. In order to prevent and resolve competition among different measurements and related practices, more cognitive alignment and agreement is needed with regard to

[12] Quote (translated from Flemish to English) derived from interview with T. Van de Graaf, Professor, Ghent Institute for International Studies, Ghent University, 13 July 2018.

[13] Interviews with F. Van der Vleuten, Senior Energy Expert, Climate Team Ministry of Foreign Affairs, The Netherlands, 5 June 2018; T. Van de Graaf, Professor, Ghent Institute for International Studies, Ghent University, 13 July 2018; and B. Hoskuldsson, Lead Partnership Specialist, Sustainable Energy for All (SEforALL), 21 September 2018.

[14] Interviews with F. Van der Vleuten, Senior Energy Expert, Climate Team Ministry of Foreign Affairs, The Netherlands, 5 June 2018; G. Fernandez Ludlow, Director for Climate Change, E. M. del Pilar Casamadrid Gutiérrez, Director for the Environment and J. Alarcón González, Head of Department for Climate Change, Ministry of Foreign Affairs, Mexico, 3 August 2018; and L. Williamson, Renewable Energy Policy Network for the 21st Century (REN21), 21 September 2018.

[15] Interview with B. Hoskuldsson, Lead Partnership Specialist, Sustainable Energy for All (SEforALL), 21 September 2018.

methodologies to determine the uptake of renewables and to find some common understanding for what constitutes a renewable source of energy.[16] Finally, it is necessary to coordinate interaction and collaboration beyond the renewable energy subfield – with sectors that deploy renewables such as transportation and heating, but also with sectors that are reluctant to deploy renewables, as well as with the issue area of energy efficiency.[17]

While these coordination efforts appear feasible, these are merely desirable as long as they do not add another level to the management structures of renewable energy institutions.[18] Furthermore, coordination attempts should neither compromise the autonomy of institutions nor constrain them in their functioning and experimenting.[19]

8.4 Assessments of Effectiveness for the Fossil Fuel Subsidy Reform Subfield

8.4.1 Assessment of Outputs, Outcomes, and Impacts

There are persuasive economic, social, and environmental reasons to tackle fossil fuel subsidies. Over the past decade, more than a dozen international institutions have begun to address this issue from various angles, from information provision and agenda setting to capacity-building and financing of reform efforts.

Taken together, these activities have led to a range of outputs. Members of several forums – including the G7 (Group of 7), G20 (Group of 20), APEC (Asia-Pacific Economic Cooperation), Friends of Fossil Fuel Subsidy Reform (Friends), and the 2030 Agenda process – have expressed commitments to phase down fossil fuel subsidies, although the precise nature of the commitment varies (Chapter 5). The G20 and APEC have also put follow-up mechanisms in place, which allow countries to report on their subsidies, and to have them reviewed by their peers (APEC Energy Working Group 2013; G20 Energy Sustainability Working Group 2013). One key requirement for reform is to ensure an adequate understanding of the scale and impacts of fossil fuel subsidies. Resources such as the Organisation for Economic Co-operation and Development (OECD) and International Energy

16 Interview with T. Van de Graaf, Professor, Ghent Institute for International Studies, Ghent University, 13 July 2018.

17 Interviews with B. Sovacool, Professor of Energy Policy, Science Policy Unit, University of Sussex, May 10, 2018; S. Röhrkasten, Scientific Project Lead Pathways to Sustainable Energy, Institute for Advanced Sustainability Studies (IASS), 17 May 2018; and L. Williamson, Outreach and Communication Manager, Renewable Energy Policy Network for the 21st century (REN21), 27 September 2018.

18 Interview with L. Williamson, Renewable Energy Policy Network for the 21st Century (REN21), 21 September 2018.

19 Interviews with B. Sovacool, Professor of Energy Policy, Science Policy Research Unit, University of Sussex, 10 May 2018; and M. Raamat, Counsellor on International Relations and Energy, Ministry of Environment, Estonia, 18 September 2018.

Agency (IEA) Inventory of Support Measures for Fossil Fuels (OECD 2018a) and the International Monetary Fund's (IMF) post-tax estimates of fossil fuel subsidies (Coady et al. 2017) seek to shed light on this question. Drawing on such work, as well as the experience of their members and external experts, institutions such as the World Bank, IMF, APEC, and Friends have supported and facilitated workshops, events, and webinars to improve governments' understanding of reform.[20]

The outputs of these various institutions can, in turn, be linked to a range of observable outcomes. Reporting mechanisms introduced under the G20 and APEC have prompted members to provide information on their domestic subsidies, although overall such estimates have been much lower than expected (Asmelash 2016), with some countries even claiming to have no subsidies at all (Van de Graaf and Blondeel 2018). Since 2015, more than a dozen G20, APEC, and Friends members have also voluntarily undergone more in-depth peer- or self-reviews, or are in the process of doing so. While the results of these exercises have in some cases been considered disappointing (e.g. ODI 2017), they have, in other cases, facilitated concrete reform plans and timelines (e.g. China 2016). Moreover, it is likely that engagement in review itself can play a valuable role in increasing internal awareness about a country's subsidies and ways to address them.[21] Interviewees have also highlighted the helpful role of workshops and other capacity-building activities to strengthen countries' understanding of fossil fuel subsidy reform (FFSR).[22]

While international institutions' activities in the FFSR space can be associated with several outputs and outcomes, it is more difficult to determine to what extent their efforts have led to increased reform on the ground. At first glance, the data on national reform activities is promising. According to the IEA, global fossil fuel subsidies dropped by just over US $300 billion between 2009 (the year the G20 and APEC committed to phase out fossil fuel subsidies) and 2015 as a result of reform (IEA 2018). The Global Subsidies Initiative estimates that around forty countries underwent some sort of FFSR between 2015 and 2017 alone (Merrill et al. 2018).

In practice, however, there is limited knowledge about the role that an institutional complex as a whole, or even an individual institution, may have in driving such reform. Indeed, it is likely that low oil prices over the past years have contributed significantly to governments' decisions to adjust or remove the subsidies they provide to consumers of fossil fuels (Benes et al. 2015): while

[20] E.g. Interview with P. G Yoshida, former Lead Shepherd, Energy Working Group, APEC, 19 July 2018; and interview with senior official, Ministry of Foreign Affairs and Trade, Government of New Zealand, 9 August 2018.

[21] Interview with R. Steenblik, former Senior Trade Policy Analyst, OECD, 20 July 2018.

[22] Interview with senior official, Ministry of Foreign Affairs and Trade, Government of New Zealand, 9 August 2018; and interview with P. G. Yoshida, former Lead Shepherd, Energy Working Group, APEC, 19 July 2018.

consumption subsidies are currently in decline (Merrill et al. 2018), upstream production subsidies appear on the rise (OECD 2018a). Domestic fiscal pressures could be another key driver of reform (Skovgaard and van Asselt 2018). Yet, while in many cases such decisions may be taken independently from the international context, global developments may also help to inform decisions taken in this regard. Steenblik, for instance, notes that a focus on tax reform in China's 2016–2020 five-year plan may have been partially informed by the country's heightened awareness of shortfalls in its internal tax expenditure monitoring system, following its FFSR peer review under the G20.[23] This may suggest that capacity-building and information-sharing can play a more significant role than high-level international commitments in driving reform. Finally, it is worth bearing in mind that efforts to promote reform may take years to come to fruition (with possible setbacks along the way). As such, this analysis should be considered an initial indication, rather than a decisive assessment, of international institutions' contributions.

8.4.2 Consequences of Institutional Complexity: What Are the Implications for Fossil Fuel Subsidy Reform?

While the overall effectiveness of international institutions in FFSR is difficult to establish, the fact that multiple international institutions are active in this area appears to be largely advantageous for promoting reform.

One advantage of institutional complexity is that different international institutions are associated with different approaches and types of expertise, such as agenda setting, capacity-building, and research.[24] The efficiency of the FFSR institutional complex is strengthened by the fact that different organizations can contribute to FFSR efforts in their speciality area(s), as opposed to one institution needing to specialize in all these approaches. At the same time, the fact that multiple institutions are engaged in this area has likely helped ensure that more resources are dedicated to international reform efforts.[25] Having multiple institutions working in this area also provides FFSR advocates with increased opportunities to keep this topic on the international agenda, with the respective framing tailored according to the institution's financial, climate change, trade, or broader social mandate.[26]

[23] Interview with R. Steenblik, former Senior Trade Policy Analyst, OECD, 20 July 2018.
[24] Interview with senior official, Ministry of Foreign Affairs and Trade, Government of New Zealand, 9 August 2018.
[25] Interview with R. Steenblik, former Senior Trade Policy Analyst, OECD, 20 July 2018.
[26] Interview with senior official, Ministry of Foreign Affairs and Trade, Government of New Zealand, 9 August 2018.

Moreover, the involvement of multiple forums allows advocates of FFSR to point to developments in other institutions to enhance ambition in their own forum, lest it falls behind.[27] One example of such positive reinforcement is the peer-review mechanisms established under both the G20 and APEC, described in more detail in Chapter 5. Nevertheless, it should be noted that such dynamics also create potential for upholding the lowest common denominator. This may constitute one explanation for why otherwise more ambitious groupings such as the Friends have opted not to set a timeline for FFSR (see Chapter 5). Finally, institutional complexity also offers the opportunity to widen the geographic scope of FFSR efforts. For instance, the fact that peer reviews are conducted under both the G20 and APEC has allowed economies from both groups to become engaged. Likewise, engagement under the High-level Political Forum on Sustainable Development (HLPF) has provided opportunities to increase traction for reform among various African countries.[28]

Despite such benefits, there are also certain challenges associated with the institutional complexity in this area. First, and like in the case of renewables, it heightens the risk of duplication of work. Although the peer reviews undertaken under the G20 and APEC can largely be seen as synergistic (Chapter 5), both forums have had to identify guidelines and approaches for review, which means that the process may have not been as efficient as it would have been under one single institution. Another, perhaps more consequential, example is the subsidy estimation work conducted by the various international organizations that produce estimates of fossil fuel subsidies, namely the IEA, IMF, OECD, and, in the future (as custodians of SDG Indicator 12.c.1), UNEP. For instance, although the IMF reports estimates for more countries than the IEA does, both organizations take a similar approach to subsidy estimates that take consumer price support as a starting point. Better coordination among the various institutions involved could help minimize the risk that labour-intensive efforts are unnecessarily repeated and, indeed, prevent inconsistencies in definitions and data from being used as an excuse to postpone action (see OECD 2018b, 10).

A second potentially problematic consequence of institutional diversity in this area has been competition among standards for defining a subsidy. At US $5.3 trillion a year, the IMF's 'post-tax' estimates of fossil fuel subsidies put such support in another order of magnitude compared to OECD and IEA estimates (Coady et al. 2017; Chapter 5).

The ambiguity has been a contributing factor for countries to claim they do not have any fossil fuel subsidies at all (e.g. South Africa, Burton et al. 2018). At the same time, such contestation of what constitutes a subsidy may have certain

[27] Ibid.
[28] Ibid.

advantages: it may enable countries to engage in reform at a pace suitable to their national circumstances;[29] and different definitions and valuation approaches can highlight complementary information about subsidies, such as, in the case of the IMF definition, their broader societal cost. The extent to which definitions diverge should also not be overstated since there are several areas of measurement on which international organizations are in agreement (Koplow 2018).

8.4.3 Management Options

These findings suggest that there are a few areas in which improved management between different institutions could enhance the effectiveness of international FFSR governance. In terms of duplication of work, while there may be an ownership rationale for the G20 and APEC to take their own approaches to VPRs, better coordination between the intergovernmental organizations involved in monitoring fossil fuel subsidies, such as the IEA, IMF, OECD, UNEP, and the World Bank, may allow both institutions to generate their data more efficiently, and indeed, is something that their members 'could be demanding'.[30] One challenge in this regard, however, is that the IEA depends on data sales for much of its revenue.

Whether international FFSR governance would benefit from enhanced coordination around the definition of a fossil fuel subsidy remains an open question. The first internationally agreed-upon methodology for the measurement of fossil fuel subsidies, issued in the context of monitoring progress on SDG indicator 12.c.1. on fossil fuel subsidies, marks an important step in this regard (UNEP et al. 2019). If efforts to elevate FFSR to the World Trade Organization (WTO) space prove successful (Verkuijl et al. 2019), the WTO's overarching definition for subsidies may also gain more traction in this context. Nevertheless, given that fossil fuel subsidies' definitional ambiguity may also be associated with certain advantages, it will be important to ensure that such a universal definition neither sets too high a threshold for action – creating a risk that some countries may disengage – nor too low a bar, which would see certain policies be needlessly excluded from FFSR discussions. Going forward, it will be important for advocates of reform to get this balance right.

8.5 Assessments of Effectiveness for the Carbon Pricing Subfield

8.5.1 Assessment of Outputs, Outcomes, and Impacts

Placing a (substantial) price on carbon emissions constitutes a key component in the response to climate change and is, according to some scholars, the most or even

[29] Ibid.
[30] Interview with R. Steenblik, former Senior Trade Policy Analyst, OECD, 20 July 2018.

the only effective instrument for mitigating climate change (Nordhaus 2008; Rabe 2018). As discussed in Chapter 6, a range of public, private, and hybrid institutions promoting carbon pricing internationally have emerged over the last twenty-five years. Due to the partial overlaps in terms of governance functions and the interlinkages between these institutions, it is difficult to pinpoint the effectiveness of the individual institutions. Yet, it is possible to discuss their combined output, outcome, and impact effectiveness.

First, outputs have been substantial and increasing exponentially over time. Companies that join private institutions – such as the UN Global Compact, Caring for Climate, or the Carbon Neutral Protocol – thereby commit to adopt mitigation strategies, including internal carbon pricing or ensuring carbon neutrality inter alia via offsetting. Joining the Carbon Pricing Leadership Coalition (CPLC), for instance, entails a commitment to support carbon pricing, although not necessarily to adopt it. When it comes to aviation, the Carbon Offset and Reduction Scheme for International Aviation (CORSIA) will in the future constitute a mandatory instrument for pricing emissions placed on the individual airlines. Regarding more technical output, most of the institutions in the subfield have been active in producing reports and other information that raised awareness and increased knowledge of carbon pricing. They also established standards for offsets and for carbon-market units. The subfield is less effective, however, when it comes to operational activities such as research and development as well as financing. Here, only the Partnership for Market Readiness provides rather modest amounts for capacity-building.

Second, concerning outcome, there has been significant behavioural change in terms of carbon-pricing policies. By early 2018, sixty-six such policies had been adopted globally by public actors (states as well as supra- and sub-national polities such as the EU or American states), and an additional fifty-two carbon pricing policies were being actively considered (Skovgaard et al. 2018).[31] The policies were put into place in all regions of the world, with a notable spike from 2015 and onwards, coinciding with the emergence of many of the institutions. While these figures do not include carbon pricing by companies on aviation, the 2016 adoption of CORSIA as a future mandatory pricing scheme in this sector fits the trend.

Yet, it is difficult to assess to what degree the output of the institutions has affected behaviour change among the actors targeted by that output (e.g. states, companies). The decision to adopt carbon pricing is inherently a polity- or business-level decision. With their standards and commitments, the governance functions of the institutions may incentivize the adoption of carbon pricing, but they may well lack the leverage to drastically alter polity- and business-level

[31] These figures do not include carbon pricing adopted by private actors.

decisions (e.g. sanctions or shaming of norm violation). From the mid-1990s to the early 2010s, private institutions, particularly IETA, have been important in promoting carbon markets, particularly in industrialized countries (Meckling 2011; Paterson 2012; Paterson et al. 2014). More recently, the Partnership for Market Readiness, as well as other institutions embedded in the World Bank and the United Nations Framework Convention on Climate Change (UNFCCC) Secretariat, have provided expert assistance to several of the developing countries that had adopted or were considering carbon pricing within the last ten years. They also facilitated learning processes among such countries.[32] It is thus plausible that this assistance has increased the likelihood of carbon-pricing proposals being adopted, but this is ultimately a subject for future research.

Third, the assessment of impact proves even more difficult. As an approximation, we briefly focus on the impact of the carbon-pricing policies that have already been implemented, while bracketing the question to which degree they have been driven by certain intergovernmental or transnational institutions. The impact effectiveness of the public carbon pricing policies can be measured in three ways. The first is to appraise the share of global emissions covered. The World Bank and Ecofys (2018) estimated that public carbon-pricing policies implemented by 2018 cover about 15 per cent of global emissions, rising to 20 per cent once the Chinese emission trading system (ETS) (scheduled for 2020) is operational. This is below the CPLC's goal of 25 per cent of global emissions by 2020, but still a sizeable share and a drastic increase compared to about 8 per cent by 2012 and less than 1 per cent by 2004 (World Bank and Ecofys 2018).

A second indicator, and arguably a more important one than the share of global emissions covered, is the incentive for reducing emissions that carbon pricing actually provides. This incentive can be measured in terms of the price of emitting one metric tonne of CO_2-equivalents. Mainstream climate economists argue that getting the price right is the chief objective of carbon pricing (Tol 2011). The price is right if it corresponds to the costs to society of emitting one tonne of CO_2 equivalents, meaning that the externality has been fully internalized. To which degree emitters actually choose to cut emissions is less important in this thinking – a notion that might be at odds with the goal to keep temperature increases to 1.5 or 2 degrees (UNFCCC 2015). Yet, no matter whether the objective is internalizing the externality of climate change or limiting temperature increases, current carbon prices are too low. According to the World Bank and Ecofys (2018), 46 per cent of the emissions subject to carbon pricing are valued below 10 US $/tonne, and most of the emissions above 10 US $/tonne are priced between 10 and 20 US $/tonne. As a reality check, the report by the High-Level Commission on Carbon Pricing

[32] Interview with senior official from the Partnership for Market Readiness, 27 August 2018.

(2017) found that price levels should be between 40 and 80 US $/tonne in 2020 to meet at least the 2 degrees target. Likewise, a meta-study of various economic estimates about the social costs of carbon found that the actual average price should be 80 US $/tonne (Tol 2011).

Third, it is difficult to isolate the impact of carbon-pricing policies, since emissions are influenced by various additional factors – including fuel prices, technology development, economic growth, and other emissions-reducing policies (e.g. fossil fuel subsidy reform or policies supporting renewable energies). These difficulties notwithstanding, existing studies of individual carbon-pricing policies indicate that their impact is limited. Carbon taxes have had at least some success in reducing emissions – as they generally have higher price levels than carbon markets and mainly have been adopted by smaller, European countries (Rabe 2018, 9). The impact of carbon markets, however, is more modest. On the upside, the sectors covered by the EU ETS, by far the world's largest carbon market, have experienced falling emissions since its inauguration in 2005. However, this drop in emissions is predominantly attributed to the economic crisis and renewable energy and energy-efficiency policies, with the ETS playing a smaller role (Hu et al. 2015; Bosello et al. 2016). Likewise, the Californian ETS, which is also among the largest carbon-pricing schemes in the world, has played a limited role in curbing emissions compared to other policy instruments such as renewable-energy policies (Bang et al. 2017).

8.5.2 Consequences of Institutional Complexity: What Are the Implications for Carbon Pricing?[33]

The carbon-pricing subfield is characterized by a number of institutions (13) that fall somewhere between the figures for the renewable energy and fossil fuel subsidy reform subfields. The same way that it was much easier to assess the output than the outcome and especially the impact or environmental effectiveness of carbon pricing, it is also easier to do so for the consequences of institutional complexity. We therefore concentrate on output and outcome in this subsection.

Regarding potential positive effects, institutional complexity has improved output in terms of simply increasing knowledge-sharing in terms of technical reports, workshops, and other dissemination formats. Concerning outcome, our informants highlight that the sheer number of institutions promoting carbon pricing should have a positive impact to keep it on the political agenda. Furthermore, the

[33] This section and the following are based on insights from interviews with officials from the UNFCCC Secretariat, the Carbon Pricing Leadership Coalition, the Networked Carbon Markets, and the Partnership for Market Readiness; as well as to a lesser degree from other carbon pricing institutions. The interviews were carried out May 2017–August 2018.

diversity of institutions, especially in terms of their memberships, helps to reach a wide variety of actors (business, policy-makers, civil society) from different sectors, e.g. with CORSIA and IATA_COP specifically targeting aviation emissions.

As for possible negative effects, the output of the various institutions is not necessarily coherent. As discussed in Chapter 6, the institutions' interpretation of the core norm of carbon pricing may diverge, with different objectives emphasized (e.g. whether the objective is functioning carbon markets or pricing emissions through carbon markets or taxes), and several informants therefore highlighted the risk of incongruent or even conflicting messages. At the same time, the interviewees also stressed that (especially cognitive) interlinkages between the institutions, as well as deliberate management efforts, help to reduce such incoherence. The risk of conflicting messages also pertains to outcome effectiveness. One example is the various institutions that promote particular standards for linking carbon markets, either under Article 6 of the Paris Agreement or for voluntary offsets. Having several standards may result in them undermining each other and incentivizing forum shopping by members. Incongruent messages also characterize the overall choice between carbon markets and taxes, since some institutions only promote carbon markets, whereas others promote both carbon markets and carbon taxes. Finally, different audiences – be they governmental, business, or civil society actors – may be less convinced if they are presented with differing arguments about the merits of carbon pricing, which may undermine the legitimacy of the instrument.

Altogether, it is difficult to gauge whether the positive consequences (larger volume of output, outreach to a diversity of audiences) outweigh the negative ones (incoherent messages and standards). This notwithstanding, and as we discuss in the following subsection, management efforts should target the negative consequences rather than reinforcing the positive ones. The positive consequences are mainly a direct consequence of the number and composition of the institutions and exist independently of management efforts.

8.5.3 Management Options

Given the current moderate levels of (mainly informal) management of the carbon-pricing subfield, it is perhaps not surprising that informants thought that more management efforts would increase effectiveness without being a game-changer. As an option, they particularly pointed to increased coordination between the existing institutions, rather than orchestration by a particular institution. Such coordination could mainly imply scaling up existing coordination efforts, while promoting a common narrative around a shared understanding of key tenets. Such tenets may include, for instance, that there is no one-size-fits-all approach but that the promotion of carbon pricing needs to be adjusted to local circumstances. With such a form of coordination, the aforementioned contradiction or incongruence of messages could be limited, as could incoherent standards for linking carbon markets or voluntary offsets.[34]

Beyond incoherence, existing informal divisions of labour, e.g. between the Partnership for Market Readiness (PMR) and the UNFCCC Secretariat (see Chapter 6), could be supplemented by a more overarching division of labour – one that also covers areas where such coordinating arrangements do not currently exist. Such arrangements are relevant for determining which institutions should target which actors, not only among specific business sectors or groups of countries but also among different ministries and agencies within the same government.[35] A respective division of labour could also take into account actors that may be currently overlooked, since institutions may easily cluster around the same actors when seeking them out independently of each other.

Our interviewees were not in agreement regarding whether coordination should cover all institutions within the subfield or just some, e.g. the UNFCCC and the World Bank–embedded institutions (CPLC, PMR, and Networked Carbon Markets, see Chapter 6). Altogether, they indicated that more coordination could improve effectiveness, mainly in terms of outcome, i.e. behavioural change, since output was already relatively high and would not be significantly affected by coordination. This said, increased coordination would not necessarily improve outcome effectiveness radically, since incoherence is not a major impediment to such effectiveness and is already addressed through existing coordination to some degree.

8.6 Conclusions

This chapter has evaluated the extent to which, and the ways in which, three subfields of the climate-energy nexus – renewables, fossil fuel subsidy reform, and carbon pricing – can be viewed as effective. The study has advanced a nuanced picture of how institutional complexity affects effectiveness, as well as presenting a set of management options for each subfield to enhance effectiveness. In what follows we compare some of the similarities and differences among the results from each subfield and discuss what our findings mean for the performance of the climate-energy nexus as a whole.

The chapter yields three broad insights. First, all three subfields appear to be successful in producing outputs, especially in terms of information-sharing and

[34] Interview with senior official from the UNFCCC Secretariat, 20 September 2018.
[35] Interview with senior official from the Partnership for Market Readiness, 27 August 2018.

capacity-building. While such outputs may seem trivial at first sight, it is important to note that institutional arrangements in other fields of global environmental governance have failed to produce significant results even at this first level of effectiveness – such as the more than 300 multi-stakeholder partnerships for sustainable development launched at the World Summit for Sustainable Development in Johannesburg 2002 (Pattberg et al. 2012).

At the outcome level, performance is more difficult to assess due to practical reasons of data-gathering and establishing causal relationships and, where this is possible, the success levels seem somewhat weaker. Results show that the institutions across the three subfields have likely influenced behavioural changes amongst actors, for example, through promoting learning processes and policy reform. However, these outcomes have not been as far-reaching as to successfully alleviate the collective action problems that are underlying the respective subfields.

On the whole, therefore, we find little concrete evidence of the institutional complexes across the subfields significantly contributing to problem solving in terms of scaling up renewables, stringent carbon pricing, and alleviation of fossil fuel subsidies. Having said that, it is unclear whether lack of outcome and impact level effectiveness is simply due to data-gathering and analytical challenges.

Second, institutional complexity can both help and hinder effectiveness. The existence of multiple institutions within a field may result in duplication of efforts and counterproductive competition, but can also create positive feedback loops and productive competition. As seen in the example of carbon pricing, even the mere fact of having many institutions working on one issue promotes political attention. In the area of fossil fuel subsidy reform, the engagement of multiple institutions has created new opportunities for FFSR advocates to keep the issue on the international agenda by framing the need for reform in different ways. A similar effect is noted in the subfield of renewable energy, which has witnessed a proliferation of actors, in particular nonstate and sub-national ones, that participate in various institutions and voluntarily create and abide to new rules and norms. To summarize these observations, the sheer magnitude of new actors and institutions that constitute the climate-energy nexus enhances political attention.

Moreover, institutional complexity may facilitate experimentation and learning at multiple venues, jurisdictions, and scales and allows for targeting actors that significantly differ in their preferences and opportunities and constraints they face, e.g. in countries with varying levels of economic and social development. This flexibility, on the other hand, risks leading to conflictive norms, forum-shopping, and diluted ambition levels. Altogether, this ambivalence is in line with previous literature on institutional complexity and fragmentation that highlights both advantages and disadvantages of such complexity (Keohane and Victor 2011; Zelli and van Asselt 2015).

Third, to the degree that international institutions can influence actors, a coordinated approach is arguably the most important factor in improving effectiveness. This should particularly imply the promotion of a common narrative and establishment of a division of labour between institutions, rather than adding further orchestration efforts for example. This said, such coordination attempts are currently largely lacking across all three subfields – which points to a major challenge: institutional complexity often arises due to a divergence of priorities amongst powerful actors (Keohane and Victor 2011), and this very conflict of interests may also hamper any coordination efforts.

In sum, institutions working within the climate-energy nexus face difficult challenges to address. On the one hand, these institutions are strengthened by having other institutions to collaborate with to reinforce their work. On the other hand, competition over resources, as well as duplications, contradictions, and incongruence of the work of different institutions, undermine some of these benefits.

The renewable energy subfield illustrates this ambiguity. On the one hand, internationally set goals provide a joint vision across institutions. On the other hand, the coordination between the multitude of institutions is far from sufficient. Further research is thus needed to look into how such coordination could be achieved to improve the effectiveness of the climate-energy nexus. Such research should look beyond the meso level and also examine interactions across subfields.

What this analysis has shown is that institutions within the three selected subfields have laid much of the groundwork for effectively contributing to their respective subfields. What is required now is the closing of governance gaps, crucially in finance and implementation, and greater cooperation between institutions to overcome some of the downsides of institutional complexity.

8.7 References

Aklin, M. J. and Urpelainen, J. 2018. *Renewables: The Politics of a Global Energy Transition*. Cambridge, MA: MIT Press.

Alcamo, J. 2017 Evaluating the Impacts of Global Environmental Assessments. *Environmental Science & Policy* 77, 268–272.

Alter, K. and Meunier, S. 2009. The Politics of International Regime Complexity. *Perspectives on Politics* 7(1), 13–24.

APEC (Asia-Pacific Economic Cooperation) Energy Working Group 2013. Guidelines on a Voluntary Peer Review for Reform of Inefficient Fossil Fuel Subsidies that Encourage Wasteful Consumption. Available at www.ewg.apec.org/documents/FINAL_VPR-IFFSR_Guidelines.pdf.

Asmelash, H. B. 2016. Falling Oil Prices and Sustainable Energy Transition. Towards a Multilateral Agreement on Fossil-Fuel Subsidies. United Nations University World Institute for Development Economics Research (WIDER) Working Paper. Available at www.wider.unu.edu/sites/default/files/wp2016-13.pdf.

Bang, G., Victor, D. G., and Andresen, S. 2017. California's Cap-and-Trade System: Diffusion and Lessons. *Global Environmental Politics* 17 3, 12–30.

Benes, K., Cheon, A., Urpelainen, J., and Yang, J. 2015. *Low Oil Prices: An Opportunity for Fuel Subsidy Reform*. New York: Columbia University Center on Global Energy Policy.

Biermann, F., Pattberg, P., van Asselt, H., and Zelli, F. 2009. The Fragmentation of Global Governance Architectures: A Framework for Analysis. *Global Environmental Politics* 9(4), 14–40.

Bosello, F., Davide, M., and Alloisio, I. 2016. *Economic Implications of EU Mitigation Policies: Domestic and International Effects*. Venice: Fondazione Eni Enrico Mattei.

Burton, J., Lott, T., and Rennkamp, B. 2018. Sustaining Carbon Lock-In: Fossil Fuel Subsidies in South Africa. In *Fossil Fuel Subsidies and Their Reform: The Politics of a Global Problem*, edited by J. Skovgaard and H. van Asselt. Cambridge: Cambridge University Press.

Chan, S., Falkner, R., Goldberg, M., et al. 2018 Effective and Geographically Balanced? An Output-Based Assessment of Non-State Climate Actions. *Climate Policy* 18(1), 24–35.

China 2016. G20 Voluntary Peer Review by China and the United States on Inefficient Fossil Fuel Subsidies that Encourage Wasteful Consumption: China Self-Review Report. Available at www.oecd.org/site/tadffss/publication/G20%20China%20Self%20Review%20on%20Fossil%20Fuel%20Subsidies-China%20Self-report-20160902_English.pdf.

Coady, D., Parry, I., Sears, L., and Shang, B. 2017. How Large Are Global Fossil Fuel Subsidies? *World Development*, 91(C), 11–27.

Easton, D. 1965. *A Systems Analysis of Political Life*. New York: Wiley.

G20 (Group of 20) Energy Sustainability Working Group. 2013. Methodology for G-20 Voluntary Peer Reviews on Inefficient Fossil Fuel Subsidies That Encourage Wasteful Consumption. Available at ru.g20russia.ru/load/783530379.

Gutner, T. and Thompson, A. 2010. The Politics of IO Performance: A Framework. *The Review of International Organizations* 5(3), 227–248.

High-Level Commission on Carbon Prices. 2017. *Report of the High-Level Commission on Carbon Prices*. Washington, DC: World Bank.

Hu, J., Crijns-Graus, W., Lam, L., and Gilbert, A. 2015. Ex-Ante Evaluation of EU ETS during 2013–2030: EU-Internal Abatement. *Energy Policy* 77, 152–163.

IEA (International Energy Agency). 2018. *World Energy Outlook 2018*. Paris: IEA/OECD. Available at www.worldenergyoutlook.org/weo2018/.

IRENA (International Renewable Energy Agency). 2018. Global Energy Transformation: A Roadmap to 2050. Abu Dhabi: IRENA. Available at www.irena.org/publications/2018/Apr/Global-Energy-Transition-A-Roadmap-to-2050.

Karlsson-Vinkhuyzen, S. I., Jollands, N., and Staudt, L. 2012. Global Governance for Sustainable Energy: The Contribution of a Global Public Goods Approach. *Ecological Economics* 83, 11-18. https://doi.org/10.1016/j.ecolecon.2012.08.009

Keohane, R. O. and Victor, D. G. 2011. The Regime Complex for Climate Change. *Perspectives on Politics* 9(1) 7–23.

Koplow, D. 2018. Defining and Measuring Fossil Fuel Subsidies. In *Fossil Fuel Subsidies and Their Reform: The Politics of a Global Problem*, edited by J. Skovgaard and H. van Asselt. Cambridge: Cambridge University Press.

Meckling, J. 2011. *Carbon Coalitions: Business, Climate Politics, and the Rise of Emissions Trading*. Cambridge, MA: MIT Press.

Meckling, J. 2019. Governing Renewables: Policy Feedback in a Global Energy Transition. *Environment and Planning C: Politics and Space* 37(2), 317–338.

Merrill, L., Gerasimchuk, I., and Sanchez, L. 2018. Fossil Fuel Subsidy Reform and Taxation, Stories for Success for the Talanoa Dialogue. IISD Global Subsidies Initiative submission to the UNFCCC Talanoa Dialogue, available at https://unfccc.int/sites/default/files/resource/61_GSI-IISD%20Fossil%20Fuel%20Subsidy%20Reform%20and%20Taxation%20Stories%20for%20Success.pdf.

Newell, P., Phillips, D., and Mulvaney, D. 2011. *Pursuing Clean Energy Equitably* Human Development Research Paper 2011/03. New York: UNDP.

Nordhaus, W. D. 2008. *A Question of Balance – Weighing the Options on Global Warming Policies*. New Haven, CT: Yale University Press.

ODI (Overseas Development Institute) 2017. Germany Ignoring Majority of Fossil Fuel Subsidies in G20 Review – ODI Experts. Available at www.odi.org/news/840-germany-ignoring-majority-fossil-fuel-subsidies-g20-review-odi-experts.

OECD (Organisation for Economic Cooperation and Development) 2018a. OECD-IEA Analysis of Fossil Fuels and Other Support. Available at www.oecd.org/site/tadffss/.

OECD (Organisation for Economic Cooperation and Development) 2018b. *OECD Companion to the Inventory of Support Measures for Fossil Fuels*. Paris: OECD.

Paterson, M. 2012. Who and What Are Carbon Markets for? Politics and the Development of Climate Policy. *Climate Policy (Earthscan)* 12(1), 82–97.

Paterson, M., Hoffmann, M., et al. 2014. The Micro Foundations of Policy Diffusion toward Complex Global Governance: An Analysis of the Transnational Carbon Emission Trading Network. *Comparative Political Studies* 47(3), 420–449.

Pattberg, P., Biermann, F., Chan, S., and Mert, A. (Eds.). 2012. *Public Private Partnerships for Sustainable Development: Emergence, Influence and Legitimacy*. Cheltenham, UK: Edward Elgar.

Rabe, B. G. 2018. *Can We Price Carbon?* Cambridge, MA: The MIT Press.

REN21. 2018. *Renewables 2018 Global Status Report*. Available at www.ren21.net/wp-content/uploads/2018/06/17-8652_GSR2018_FullReport_web_final_.pdf.

Röhrkasten, S. 2015. *Global Governance on Renewable Energy - Contrasting the Ideas of the German and Brazilian Governments*. Wiesbaden: Springer VS.

Röhrkasten, S. and Westphal, K. 2013. *IRENA and Germany's Foreign Renewable Energy Policy - Aiming at Multilevel Governance and an Internationalization of the Energiewende?* Berlin: SWP.

Skovgaard, J. and van Asselt, H. 2018. The Politics of Fossil Fuel Subsidies and Their Reform: An Introduction. In *Fossil Fuel Subsidies and Their Reform: The Politics of a Global Problem*, edited by J. Skovgaard and H. van Asselt. Cambridge: Cambridge University Press, 3–20.

Szulecki K., Pattberg, P., and Biermann F. 2011. Explaining Variation in the Effectiveness of Transnational Energy Partnerships. *Governance* 23(7), 713–736.

Tallberg J., Sommerer T., Squatrito, T., and Lundgren M. 2016. The Performance of International Organizations: An Output-Based Approach. *Journal of European Public Policy* 23(7), 1077–1096.

Tol, R. S. J. 2011. The Social Cost of Carbon. *Annual Review of Resource Economics* 3, 419–443.

Underdal, A. 2002. One Question, Two Answers. In *Environmental Regime Effectiveness: Confronting Theory with Evidence*, edited by E. L. Miles, A. Underdal, S. Andresen, J. Wettestad, J. B. Skjærseth, and E. M. Carlin. Cambridge, MA: MIT Press.

UNEP (United Nations Environment Programme), OECD (Organisation for Economic Co-operation and Development), and IISD (International Institute for Sustainable

Development). 2019. *Measuring Fossil Fuel Subsidies in the Context of the Sustainable Development Goals*. Nairobi: UNEP.

UNFCCC (United Nations Framework Convention on Climate Change). 2015. *The Paris Agreement*.

United Nations. 2018. *The Sustainable Development Goals Report 2018*. New York. https://doi.org/10.18356/3405d09f-en.

Urpelainen, J. and Van de Graaf, T. 2015. The International Renewable Energy Agency: A Success Story in Institutional Innovation? *International Environmental Agreements: Politics, Law and Economics* 15(2), 159–177. https://doi.org/10.1007/s10784-013-9226-1.

Van de Graaf, T. and Blondeel, M. 2018. Fossil Fuel Subsidy Reform: An International Norm Perspective. In *Fossil Fuel Subsidies and Their Reform: The Politics of a Global Problem*, edited by J. Skovgaard and H. van Asselt. Cambridge: Cambridge University Press.

Van de Graaf, T. and Zelli, F. 2016. Actors, Institutions and Frames in Global Energy Politics. In *The Palgrave Handbook of the International Political Economy of Energy*, edited by T. Van de Graaf, B. K. Sovacool, A. Ghosh, F. Kern, and M. T. Klare. London: Palgrave Macmillan.

Verkuijl, C., van Asselt, H., Moerenhout, T., Casier, L., and Wooders, P. 2019. Tackling Fossil Fuel Subsidies through International Trade Agreements: Taking Stock, Looking Forward. *Virginia Journal of International Law* 58(2), 309–368.

World Bank and Ecofys. 2018. *State and Trends of Carbon Pricing 2018*. Washington, DC: World Bank.

Young, O. R. 2002. *The Institutional Dimensions of Environmental Change: Fit, Interplay, and Scale*. Cambridge, MA: MIT Press

Zelli, F. and van Asselt, H. 2015. Fragmentation. In *Research Handbook on Climate Governance*, edited by K. Bäckstrand and E. Lövbrand. Cheltenham, UK: Edward Elgar.

9

Conclusions

Coherence, Management, Legitimacy, and Effectiveness in the Climate-Energy Nexus

FARIBORZ ZELLI, KARIN BÄCKSTRAND, NAGHMEH NASIRITOUSI, JAKOB SKOVGAARD, AND OSCAR WIDERBERG

Combating climate change and transitioning to fossil-free energy systems are two central planetary challenges humanity faces today. The two challenges strongly overlap in their substance and the political choices we have to address them. As a consequence, a plethora of international institutions – public, private, and hybrid ones – fall right into this overlap. They seek to regulate the many complex linkages between climate change and decarbonized energy systems.

The introduction to this volume stressed the urgency of researching these links at the nexus of global climate change governance and energy governance. The highly complex and fragmented nature of the climate-energy governance nexus confronts us with a series of questions that need timely responses. How coherent is this governance nexus – in terms of the memberships, functions, core norms, and interactions across its numerous institutions? How have states, international organizations, and other actors responded to this complexity when they sought to tackle coherence and governance gaps? How does institutional complexity affect the legitimacy and effectiveness of individual institutions and the governance of the climate-energy nexus as a whole? Or, to concentrate these concerns into one question: to which degree do the many governance efforts on climate change and energy transition today add up to a coherent regulatory and institutional global framework?

It has been the purpose of this book to provide first and crucial answers to these questions and to alert a wider audience to their importance. In Section 9.1, the concluding chapter summarizes some of the novel findings from our research. The summary is by no means exhaustive, and we refer the readers back to the previous chapters for the rich, cross-dimensional, and comparative insights they provide. Our research findings show us in great detail what is at stake and how carefully institutions and stakeholders need to navigate in the complexity of the climate-energy nexus with its many and emergent governance arrangements. Against this backdrop, this chapter also discusses possible explanations of our results

(Section 9.2) and derives policy recommendations from them (Section 9.3). It also points to some of the next steps we need to take as researchers and stakeholders in order to live up to the tall order of tackling climate change and energy transition simultaneously (Section 9.4).

9.1 Findings

This section is organized along the main components of the analytical framework introduced in Chapter 2. The major backbone of the framework are four evaluative themes or analytical dimensions to grasp the shape and performance of institutions in complex governance systems: coherence, management, legitimacy, and effectiveness. Each of these evaluative themes was operationalized through particular dimensions.

To examine the four themes, the framework combined both rationalist and constructivist accounts derived from theories of International Relations. Coherence and management were to a large extent scrutinized along organizational dimensions such as institutional membership and governance functions. The dimension of the core norm, however, added an important constructivist dimension to the analysis of coherence. For the two themes that concern performances under conditions of complexity, i.e. legitimacy and effectiveness, the authors mostly relied on a constructivist or sociological perspective. They examined perceptions and expectations by selected audiences and stakeholders involved in international negotiations on energy and climate.

In addition to this dual approach, the framework distinguished three analytical levels to get a more encompassing grasp on the climate-energy nexus: the macro level (the climate-energy nexus as a whole); the meso level (the institutional complexes that govern each of the nexus's subfields analyzed in this volume, namely renewable energy, fossil fuel subsidy reform [FFSR], and carbon pricing); and the micro level (specific interlinkages between individual institutions within each of these subfields).

9.1.1 Coherence

Coherence was defined in Chapter 2 as the degree of harmony or compatibility of institutional features to one another or to an overarching purpose. This theme was studied along four dimensions: institutional membership, coverage of governance functions, adherence to an overarching core norm, and interaction mechanisms. Two of these dimensions (membership and functions) were analyzed across all three analytical levels. The other two were only scrutinized for the meso level (core norm) and micro level (interaction mechanisms), respectively.

In Chapter 3, Sanderink and colleagues advanced a novel and unique dataset. They mapped the institutions across the entire climate-energy governance nexus at the macro level and examined in great detail the *membership* distribution of public, civil society, and firm-based members. The database they generated contains 108 institutions, which include more than 12,000 members as of early 2017. Through their network analysis, the authors identified the International Renewable Energy Agency (IRENA) and the United Nations Framework Convention on Climate Change (UNFCCC) as the central institutions within the climate-energy nexus, due to their high degrees of universality and inclusiveness.

By the same token, Sanderink and colleagues found for the climate-energy nexus a clear dominance of public members, i.e. country, state, or municipal governments, secretariats to international agreements, and other public agencies. These are involved in no less than seventy-eight institutions and exclusively constitute forty-eight of these, with cities being by far the most frequent type of public member. While the predominant number of public members come from developed countries, public actors from various countries in the global South, such as China, Mexico, and Indonesia, have become widely represented across the nexus. By contrast, firm or business actors are members of less than half of the sample (fifty-one institutions), and only make up seventeen institutions exclusively. Civil society actors are officially engaged in solely thirty-five institutions, seven thereof being purely civil-society-based.

When disentangling these overall observations for the three selected policy subfields, the contributions to this volume equally identified a predominance of public actors, albeit with certain variations across issue areas. As Sanderink demonstrated in Chapter 4, the renewable energy subfield of the nexus exhibits the most institutions of the three cases, forty-six in total. The membership distribution across these institutions mirrors that of the climate-governance nexus as a whole. Twenty-eight of the forty-six renewable energy institutions are solely constituted by public members. Apart from the UNFCCC and IRENA, these include, for instance, the International Energy Agency (IEA), the Association of Southeast Asian Nations, and the Covenant of Mayors. No more than four institutions are merely firm-based and only three are purely civil-society-based, with the remaining eleven institutions featuring mixed memberships.

For fossil fuel subsidy reform, Verkuijl and van Asselt in Chapter 5 identified an even greater imbalance in favour of public actors. The institutional complex is made up of only fourteen institutions, thirteen of them purely intergovernmental, e.g. the UNFCCC, IEA, the United Nations Environment Programme (UNEP), the Friends of Fossil Fuel Subsidy Reform (Friends), and the World Bank. The only exception is the Global Subsidies Initiative, which was started by a Canadian NGO.

By contrast, Skovgaard and Canavan found a nearly balanced distribution for the subfield of carbon pricing (Chapter 6). Of the thirteen institutions they identified for this complex, five were solely public (e.g. the UNFCCC or the International Carbon Action Partnership [ICAP]), four firm-based (e.g. the International Emissions Trading Association [IETA] and the Verified Carbon Standard), and one, the Gold Standard, constituted by civil society actors, with the three remaining ones being hybrid institutions.

For the second dimension of coherence, *governance functions,* Chapters 2 and 3 differentiated four major dimensions to guide the empirical analyses in this volume: standards and commitments, operational activities, information and networking, and financing. At the macro level of the climate-energy nexus as a whole, Sanderink and colleagues (Chapter 3) found that all these four governance functions were institutionally covered. This notwithstanding, there was no clear division of labour evolving across the nexus as a whole. Instead, they identified a rather uneven distribution of functions. The climate-energy nexus is dominated by institutions that perform information and networking functions, while, for instance, standard-setting and financing are conducted by a much smaller set of institutions. Institutions with public members are clearly in the majority when it comes to information and networking functions. Interestingly, though, the number of public and firm-based institutions is nearly equal for the setting of standards and commitments. Put differently, a key observation is that soft-governance functions – such as information and networking – dominate at the expense of hard-governance functions that involve authoritative rule and standard-setting. Clearly, states are not willing to cede sovereign control of rule-setting on energy to global institutions.

When zooming in on the meso level and comparing the three selected cases, this picture changes and gets more differentiated. The subfield of renewable energy again comes closest to the research findings from the climate-energy nexus as a whole. In Chapter 4, no clear cross-institutional division of labour of governance functions can be found. Instead, global renewable energy governance is geared toward information and networking functions that are dominated by public institutions.

For the subfield of fossil fuel subsidy reform, by contrast, the distribution of institutions across governance functions is much more balanced. As Verkuijl and van Asselt observe in Chapter 5, this is partly the result of active coordination by international organizations and country governments. Some forums are instrumental in agenda-setting and the formulation of commitments (e.g. the Group of Twenty [G20] and Friends), while others concentrate on the provision of information (e.g. the Organisation for Economic Co-operation and Development [OECD] and the IEA) or on financing and implementation on the ground (e.g. the World Bank and the Global Subsidies Initiative).

For the third case study on carbon pricing, Skovgaard and Canavan established yet a different picture, namely one of considerable overlaps over certain governance functions. For instance, ICAP (intergovernmental), IETA (business), and the Networked Carbon Markets Initiative (NCM; hybrid) engage in very similar information and networking activities to promote the linking of carbon markets. However, in all subfields rule- and standard-setting is a less predominant governance function, which reflects that sovereign states have a solid grip on decision-making power in energy governance.

As a third dimension to analyze the degree of institutional coherence in the nexus, Chapter 2 identified the *core norm*, i.e. the overarching expectation of appropriate behaviour that characterizes a particular subfield. The concept is closely related to the very definition and delineation of a subfield and its respective institutional complex. It provides the substance or goal that unites the institutions governing the area in question. Although the institutions agree on the importance of the norm, they may interpret its precise content and application in strongly diverging ways (see also Wiener 2004). The interpretation of the core norm therefore served as the major benchmark for identifying normative convergence or divergence at the meso level.

For renewable energy, Sanderink (Chapter 4) identified Sustainable Development Goal 7.2 as the core norm, which aspires to substantially increase the share of renewable energy in the global energy mix, especially with a view to ensuring access to, and availability of, clean energy for all (United Nations 2015). The universal nature of this norm notwithstanding, Sanderink ascertained a considerable normative divergence and contestation across the subfield. Only seventeen of forty-six institutions subscribe to the norm literally, while the others prioritize specific aspects over others, i.e. either climate change mitigation, energy access, or energy security.

For fossil fuel subsidy reform, Verkuijl and van Asselt (Chapter 5) found a similar, if not even more divergent picture. They also referred to a key institutional document as the most suitable expression of the core norm, in this case taken from the statement of the third leaders' summit of the G20 in Pittsburgh 2009: to '[r]ationalize and phase out over the medium term inefficient fossil fuel subsidies that encourage wasteful consumption' (G20 2009, paragraph 29). Verkuijl and van Asselt caution that the key terms in this formula remain contested across actors and institutions, including what should be counted as fossil fuel subsidies in the first place. Moreover, phase-out dates remain ambiguous across the institutional complex. In addition, the United States under the Trump administration has begun to voice overall reservations on the FFSR norm in general. Against this backdrop, it is not surprising that the authors found no major institution to which all others would normatively adhere to.

Normative divergence and contestation, rather than convergence, also mark the institutional complex on carbon pricing (Chapter 6). As a core norm for this subfield, Skovgaard and Canavan established that climate change is best mitigated by giving a price signal to emitters and by leaving the decision of how to reduce carbon emissions to the market. This said, they identify no less than four aspects along which the interpretation of this norm considerably varies across the complex. These four aspects are: quantity versus price instruments, mandatory versus voluntary schemes, carbon pricing within a given jurisdiction versus off-setting, and whether polluters should pay for all of their emissions or not. Variation in institutional positions concerning these four aspects creates normative clusters of largely private institutions on one side and hybrid or public ones on the other.

The fourth and final dimension of coherence introduced in Chapter 2 regards the interaction *mechanisms* at the micro level of individual institutions. Three types of mechanisms were distinguished through which individual institutions may have an impact on each other within the climate-energy nexus: cognitively, through a flow of knowledge and information; normatively, through an imitation of, or adaptation toward, norms and rules; or behavioural, through the functional or strategic behaviour of specific members or other actors.

Chapters 4–6 found plenty of evidence for all three mechanisms being at play in the three institutional complexes, with no particular dominance of one over the others. Yet, given the aforementioned overall dominance of information and networking functions in the climate-energy nexus, it was arguably easier to identify cognitive interactions than normative interactions. Behavioural interactions depend in part on the overlaps of memberships across institutions. To give one example: in the institutional complex on FFSR (Chapter 6), their membership in both the G20 and the Asia Pacific Economic Cooperation (APEC) allowed the United States and China to push successfully for reform commitments and peer-review efforts in both intergovernmental institutions. Such reform efforts included events and initiatives linked to the UNFCCC, the High-Level Political Forum on Sustainable Development, and the World Trade Organization (WTO). This form of forum shopping (cf. Raustiala and Victor 2004; Orsini et al. 2013) has played an important role in increasing the saliency of FFSR on the international policy agenda.

9.1.2 Management

The second evaluative theme, management, was specifically targeting the micro level, i.e. relations among individual institutions within a given subfield of the climate-energy nexus. Quoting Stokke (2001, 11), Chapter 2 defined management as any deliberate efforts taken 'by participants in tributary or recipient regimes to prevent, encourage, or shape the way one regime affects problem solving under another'. The chapter distinguished two main dimensions for assessing such efforts. First, the levels (regional, national, global) and agents (domestic or sub-domestic actors, one or several affected institutions, or an overarching institutions) of management; and second, the actual consequences and potential successes of management efforts. In other words: did the management attempts yield any convergence about the core norm or entail any sensible and efficient distribution of memberships and governance functions across the affected institutions?

Notably, all three case studies identified a variety of effective management attempts for the particular institutional interlinkages they put under scrutiny. This signals a strong awareness by institutional members and third parties of the complexity in which they are operating. In Chapter 4 on renewable energy, Sanderink concentrated her micro-level analysis on three hybrid institutions, i.e. partnerships made up by all three major stakeholder types (public, firm-based, civil society). For all three of them – the Renewable Energy and Energy Efficiency Partnership, The Renewable Energy Policy Network for the 21st Century (REN21), and Sustainable Energy for All – she found management attempts initiated by different drivers at various levels. For example, joint coordination efforts between the three institutions included a shared portal and a joint report for common information. Further convergence was reached through unilateral management approaches, with each partnership reaching out toward important third institutions, e.g. IRENA. Finally, Sanderink identified the UNFCCC and Agenda 2030 as overarching institutions that provide core global goals upon which the three partnerships converge. These diverse management activities notwithstanding, she stressed an important limitation of her micro-level study. In light of a total of forty-six institutions and a dominance of public institutions in the renewable energy subfield, the relations between the three hybrid institutions selected should not be mistaken as representative of the management attempts in this institutional complex. In fact, Sanderink found that, aside from the three partnerships that she examined, only thirteen other institutions explicitly link their activities to the UNFCCC and Sustainable Development Goal (SDG) 7.

The results of the micro-level studies were slightly more representative for the two other case studies, FFSR and carbon pricing, since these two subfields exhibit a much lower number of institutions. For the institutional complex on FFSR (Chapter 5), consisting of but fourteen institutions, Verkuijl and van Asselt studied three key intergovernmental institutions in depth: G20, APEC, and Friends. They largely identified informal management efforts between these three and their members, taking the form of meetings on the side during international events, outreach to Friends members for their expertise, and outreach from Friends members to other countries in advance of G20 summits. Despite this informal character, or likely because of it, the management attempts were very effective, since they facilitated the information flow and kept the level of ambition high across the three institutions. Friends members in particular adopted a broker role and proactively enhanced complementarity among the three institutions. For instance, developments at the G20 gave New Zealand, a Friends member, leverage to ensure that similar efforts were undertaken under APEC, in terms of both reform commitments and the introduction of a peer review process in this area.

For their micro-level examination on carbon pricing in Chapter 6, Skovgaard and Canavan arguably chose the most powerful institutions among the thirteen that constitute the meso-level institutional complex. When analyzing interactions between the UNFCCC and World Bank–embedded institutions, they determined various joint management efforts. Such efforts mainly took place through regular, yet often informal, contacts and meetings between institutional officials that are specialized in carbon pricing questions. These experts helped to defuse situations that could potentially have led to open competition or conflict. On the other hand, and unlike for renewable energy and FFSR, Skovgaard and Canavan did not find any noteworthy unilateral or overarching attempts by particular states or institutions to manage carbon pricing. Moreover, most management attempts for this policy field have originated in an ad hoc or incremental manner rather than going back to deliberate and long-sighted planning. On balance, the institutional sample for carbon pricing thus features less encompassing management efforts than the samples in the other two case studies.

Table 9.1 summarizes the major findings for the two evaluative themes of coherence, and management across the three institutional complexes for renewable energy, FFSR, and carbon pricing. The left column lists the various dimensions, while the three columns on the case studies clarify to which extent these dimensions were addressed in a coherent manner, i.e. whether or not the three subfields exhibit high and balanced scope of memberships, functions, interaction mechanisms, a convergence on core norms, and a high and effective scope of management efforts. The bottom row includes our overall assessments of the shape and direction of the institutional complexes for each subfield. The terminology for this overall assessment (here: coordination for all three cases) is guided by the book's analytical framework (see Chapter 2, Table 2.1) that distinguishes largely non-managed from largely managed institutional constellations under different degrees of coherence.

Our findings indicate that all three subfields are characterized by coordination, rather than competition or outright harmony. This reflects the aforementioned results for coherence and management. On the one hand, all three policy subfields exhibit a medium degree of normative and functional convergence. The institutions in each of them share a certain core norm, for example phasing out inefficient fossil fuel subsidies, but differ considerably in their interpretations of that core norm. While all institutions in each subfield cover all governance functions, they are frequently skewed toward information and networking at the expense of standard-setting, and

Table 9.1 *Findings for coherence and management in the climate-energy nexus.*

	Nexus Subfields		
Evaluative Themes	Renewable Energy	Fossil Fuel Subsidy Reform	Carbon Pricing
Coherence			
Scope and distribution of memberships (meso)	46 institutions Imbalance, dominance of public institutions (28)	14 institutions Imbalance, strong dominance of public institutions (13)	13 institutions Balanced distribution (5 public, 4 firm-based, 1 civil society-based)
Scope and distribution of governance functions (meso)	High Imbalanced distribution in favour of information & networking	High Balanced and complementary distribution	High Balanced, but competitive distribution
Convergence on core norm (meso)	Low to medium Strong variations across goal priorities	Low to medium Strong variations on the core definition of fossil fuel subsidies	Medium Strong variations across goal priorities and approaches to reach these goals
Scope and distribution of interaction mechanisms (micro)	High All types (cognitive, normative, behavioural)	High All types (cognitive, normative, behavioural)	High All types (cognitive, normative, behavioural)
Management			
Scope of efforts across agents and levels (micro)	High Both formal and informal Largely joint and unilateral, partly overarching	High Mostly informal Joint, unilateral and overarching	Medium Both formal and informal, but largely ad hoc Largely joint, no major unilateral or overarching efforts
Consequences (micro)	Facilitating convergence	Facilitating convergence	Preventing competition
Overall Assessment	*Coordination*	*Coordination*, evolving *division of labour*	*Coordination*

public actors are over-represented compared to private stakeholders. On the other hand, all three cases exhibit diverse and relatively successful management attempts across key institutions, with a certain caveat regarding renewable energy where more studies are needed to scope out this relatively large institutional complex.

The combination of these two qualities – medium coherence and advanced management – suggests that the meso-level institutional complexes on renewable energy, FFSR, and carbon pricing fulfil notions of 'coordination'. The differences across the three subfields are, hence, rather minor. FFSR comes closest to a division of labour, due to the widespread management efforts amongst a small number of institutions and the balanced distribution of governance functions across these institutions. The institutional complex on carbon pricing, on the other hand, depends upon ad-hoc management efforts to prevent outright competition or conflict across institutions.

9.1.3 Legitimacy

Moving from the shape or degree of institutional complexity to its consequences, the remaining two evaluative themes of legitimacy and effectiveness have been conceptualized in sociological terms, i.e. they are measured by perceptions among key audiences and stakeholders. To this end, the analyses relied on interviews and questionnaires to explore how certain actors perceive the legitimacy and effectiveness of selected institutions in the three policy areas that we put under scrutiny in this volume.

To examine consequences of institutional complexity for legitimacy, Chapters 2 and 7 introduced a total of nine dimensions. The dimensions capture criteria from the diverse scholarship on legitimacy concepts (normative and sociological, input and output legitimacy), that were geared to be analyzed in terms of perceptions among key stakeholders. They include five dimensions of input legitimacy (an institution's expertise, inclusiveness, procedural fairness, transparency, and accountability) and four dimensions of output legitimacy (produced output, behaviour-changing outcome, problem-solving impact, and distributive fairness).

In Chapter 7, Nasiritousi and Verhaegen advanced this novel framework, guided by theoretical assumptions and expectations on the legitimacy perceptions of these dimensions (see also Section 9.2). The framework was employed to empirically examine five selected institutions in the subfield of renewable energy: the UNFCCC, IEA, IRENA, the Clean Energy Ministerial (CEM), and REN21. To get reliable results from a broad set of stakeholders, an expert survey was conducted with 262 respondents, and analysed by performing an exploratory factor analysis for the 9 dimensions. The findings are presented and visualized in detail in Chapter 7, differentiated by different stakeholders and institutions.

We can only highlight some of the general findings on legitimacy perceptions among key stakeholders for the renewable energy subfield. One core finding is that the UNFCCC was regarded as the most legitimate institution in the sample, while the Clean Energy Ministerial (CEM) received the least positive legitimacy assessment across the nine dimensions, with a noteworthy margin to the other four institutions. Moreover, the UNFCCC scored higher compared to any other institution with regard to dimensions of input legitimacy. An explanation can be that the UNFCCC is the only multilateral institution with a political mandate in the institutional complex on renewable energy, and thereby has the most elaborated procedures to promote access, inclusiveness, and transparency in particular. Apart from these observations, however, it is striking how similar the overall legitimacy assessments by key stakeholders were for the five institutions. Importantly, this suggests that stakeholders do not adequately disentangle their views of individual institutions from their overall impression of the renewable energy subfield – at least not for those institutions that have similar functions, feature overlapping mandates, and regularly engage in interactions and management efforts.

When distinguishing the results according to different subsets of stakeholders, though, Nasiritousi and Verhaegen found that an institution's formal fulfilment of normative legitimacy criteria does not necessarily translate into positive legitimacy assessments by all actors. Rather, the varying specializations and backgrounds of stakeholders led to significant variations in legitimacy assessments – owing to differences in their normative orientations, value sets, and familiarity with certain institutions. In this context, the professional background was much more significant for legitimacy assessments than the geographical origins of respondents, which on balance did not render any significant differences.

9.1.4 Effectiveness

Three dimensions of effectiveness were differentiated in Chapter 2, and applied in Chapter 8: the institutions' production of information, norms, and policies (output), the effect on institutional members and their behaviour (outcome), and the ultimate effect on solving the climate- or energy-related problem that the respective institutions sought to address (impact) (cf. Underdal 2002; Tallberg et al. 2016).

These were identical with the three-tiered legitimacy dimensions discussed in the previous section on legitimacy. What set the study in Chapter 8 apart from that in Chapter 7 was, first, that a broader sample was analyzed, namely across all three institutional complexes – renewable energy, fossil fuel subsidy reform, and carbon pricing. Second, the empirical examination of effectiveness combined both document analyses and qualitative data on the perceptions and expectations of experts (derived from interviews).

A core finding was that all three institutional complexes are relatively successful in producing outputs, especially in terms of information-sharing and capacity-building. Achievements on the outcome and impact levels were harder to establish, but the overall evidence suggested that none of the institutional complexes has been capable of substantially mitigating the collective action problems they set out to address. Not surprisingly, the interviewees called for more coordinated approaches for all three issues to build common narratives and promote a certain division of labour across institutions.

Disentangling these general insights for each of the three subfields showed that institutional complexity can be a supportive as well as hindering factor for effectiveness in the climate-energy nexus. For renewable energy, the interviewees expected more advantages than disadvantages from the institutional complexity in this subfield. This goes in particular for the output level, with benefits arising from multiple perspectives and knowledge bases and thanks to a pragmatic compartmentalizing approach to master different challenges of the energy transition. The most frequently named disadvantages, on the other hand, included: duplication of work; competition over resources, influence, and visibility; and the lack of an umbrella institution to address these issues.

A similarly positive assessment of effectiveness by core audiences was established for the institutional complex on fossil fuel subsidy reform. Similar to renewable energy, respondents viewed the availability of different types of expertise and experiences across the various institutions as highly beneficial, especially for output effectiveness. With regard to outcome, the different memberships of institutions working on FFSR helped to widen the geographical spread of governance efforts on this issue. In addition, interviewees welcomed a race to the top or positive reinforcement arising from institutional plurality, i.e. the incentive to keep or surpass the level of ambition of other institutions in the field. The disadvantages that were most frequently mentioned coincided with those for renewable energy, namely duplication of work and competition among standards at the output level, implying considerable transaction costs and ambiguity regarding outcome effectiveness.

These arguments about the consequences of institutional complexity were more or less repeated for carbon pricing. On the positive side, respondents stressed the knowledge distribution for output effectiveness, as well as positive reinforcement effects and stronger outreach activities for outcome effectiveness. Disadvantages that the interviewees referred to were equally similar to the other two subfields: the competition among standards and the conflicting messages this sends to stakeholders. Importantly, and in comparison to renewable energy and FFSR, these negative implications for carbon pricing were highlighted quite frequently by respondents. This confirmed the aforementioned findings on coherence and management for this subfield, which suggested a somewhat more competitive nature of the institutional complex. The findings for effectiveness echoed this notion and suggest that the degree and consequences of this competitiveness merit further exploration.

9.2 Explanations

As laid out in Chapter 1, this edited volume first and foremost has sought to provide new conceptual and empirical insights into the shapes and consequences of institutional complexity for the nexus of two major policy domains, climate change and energy. Given the lack of previous comparative studies about the climate-energy nexus, this book combined a series of innovative methods and research steps, which, in turn, generated a large set of important and novel research findings for scholars and practitioners alike. In the following, we briefly focus on possible explanations for these findings on the four evaluative themes of coherence, management, legitimacy, and effectiveness.

9.2.1 Explaining Coherence and Management in the Climate-Energy Nexus

Chapter 1 briefly proposed one explanation for variation across the subfields on coherence and management, namely the position of each subfield within the climate-energy nexus. The variation in this position also served as one of our main criteria for selecting the three case studies for this volume. Carbon pricing is mainly connected to climate change, with mitigation as its core objective. Renewable energy, on the other hand, is more at the heart of energy governance, as a policy subfield primarily seeking to promote a segment of energy sources, while mitigating climate change comes as a co-benefit or secondary goal. Fossil fuel subsidy reform, finally, falls somewhere in the middle of the two other subfields, as it is both an instrument for climate mitigation and for de-carbonization of the energy mix.

In Chapter 1, we expected that these different positions of our three case studies within the climate-energy nexus would matter for the coherence of the respective institutional complexes – especially since climate change governance was, unlike energy governance, marked by a central institution, the UNFCCC, as a hub of multilateral climate diplomacy (Biermann et al. 2009; Van de Graaf and Colgan 2016). This central position, and possibly coordinative influence of the UNFCCC, would make it well-equipped to play a stronger role for predominantly climate-related topics such as carbon pricing.

However, when looking at the findings for coherence and management, as displayed in Table 9.1, these expectations were not supported by the empirical evidence. Instead, in spite of their very different positioning in the climate-energy nexus, all three institutional complexes share a number of institutional features, such as wide coverage of different governance functions, dominance of public institutions (albeit to different degrees), and low or medium convergence on core norms, as well as multiple and effective management attempts. Consequently, they score similarly in the overall assessment of 'coordination', i.e. they exhibit a relatively well managed, medium level of coherence.

What is more, the convening, steering or orchestrating role that the UNFCCC plays in global climate governance in general did not have the expected impact for the subfield of carbon pricing. On the contrary, Skovgaard and Canavan (Chapter 6) found no evidence for any management attempts by the UNFCCC for this subfield. Rather, as Sanderink showed in Chapter 4, the UNFCCC provides such an umbrella function for some of the institutions governing renewable energy.

At second glance, however, the positioning in the climate-energy nexus appears to play some role for shaping the institutional complexes, albeit not the only or major role. For instance, the normative guidance that the UNFCCC provides for the renewable energy subfield affects a minority of institutions in the complex. Sanderink's findings point to roughly one-third of institutions that adhere explicitly to core norms of the UN climate regime. By contrast, output from the UNFCCC, particularly the Kyoto Protocol and the Paris Agreement, have historically often acted as stimuli for scaling up efforts across the carbon pricing subfield.

This said, the potential relevance of the nexus position for the degree of institutional complexity may be considerably qualified by other factors. In the case of carbon pricing, for instance, one reason for the slightly more competitive nature of the complex (e.g. with regard to the distribution of governance functions) is the constellation between the UNFCCC on the one hand and World Bank–embedded institutions on the other. These two sets of international institutions differ, for instance, in their confidence in the ability of the market to yield carbon-emission reductions (Chapter 6). While this constellation confirms the important role that the UNFCCC has for this subfield, it shows that, for a more differentiated explanation, other phenomena have to be taken into account.

Concretely, the climate-energy nexus does not exist in a vacuum, but overlaps with other policy domains with their own institutional settings. Carbon pricing is not only a matter of climate change and energy, but also of financing and international development, which explains the significant role the World Bank plays in balancing the UNFCCC's dominant position in this subfield. Likewise, renewable energy is linked to a plethora of economic, environmental, and social concerns that are in part regulated by other institutions. It is hence not surprising that renewable energy is governed by the biggest institutional complex in our sample. This high number of institutions, in turn, implies both challenges and opportunities that may impact the institutional features of the complex, from more options to cover governance functions on the one hand to more difficulties to reach overarching coordination on the other (Scott 2008; Biermann et al. 2009).

These considerations about thematic scope and the number of institutions demonstrate the need for more theory development in order to explain and understand the variation of institutional features in a governance nexus. This development can build, inter alia, on different theories on international institutions who continuously incorporated the phenomenon of institutional complexity over the past two decades. Rationalist, sociological and discursive accounts of institutionalism yield different theoretical expectations that could be adapted to the study of a governance nexus such as the one on climate and energy. Among the rationalist approaches, proponents of instrumental multilateralism could, for instance, analyze the role of hegemonic countries in determining which core rules are adopted, which governance functions are covered and which management attempts are taken by which actors or institutions (cf. Ikenberry 2003; Morse and Keohane 2014). Likewise, neoliberal institutionalists and organizational ecologists could scrutinize the role of underlying constellations of interests or situation structures (cf. Zürn 1993; Keohane and Victor 2011), problem structures (cf. Rittberger and Zürn 1990; Underdal 2002; Zelli et al. 2017), or resource dependencies (cf. Abbott et al. 2016) in shaping certain areas of a governance nexus. Sociological and discursive institutionalists, finally, could help us understand to what extent the (lack of) institutional coherence in a nexus represents overarching norms or discourses – and the rivalries and contestations among them – in which the various institutions are embedded (cf. Conca 2006; Schmidt 2008, 2017; Arts and Buizer 2009).

9.2.2 Explaining Legitimacy and Effectiveness in the Climate-Energy Nexus

In Chapter 7, Nasiritousi and Verhaegen examined a variety of potential reasons for the legitimacy assessments by stakeholders. They followed Steven Bernstein and other scholars who have shown that different types of stakeholders may hold different legitimacy demands based on their social values, norms, and previous experiences (Bernstein 2005; Karlsson-Vinkhuyzen and Vihma 2009; Lenz and Viola 2017). Nasiritousi and Verhaegen therefore examined whether varying stakeholder characteristics made a difference for legitimacy assessments under conditions of institutional complexity. For their sample of five renewable energy institutions, they found that some of these characteristics indeed mattered for legitimacy assessments. These assessments particularly varied between governmental and nonstate actors and between stakeholders with different work profiles and orientations. The geographic origin, on the other hand, played no major role for how the legitimacy of an institution was valued. These results indicate that norms, values, and experiences of audiences may be of relevance for the sociological legitimacy of institutions in a nexus.

In addition, Nasiritousi and Verhaegen took into account the degree of institutional complexity as a potential explanation for shaping audiences' legitimacy beliefs. Following cognitivist assumptions about limitations of visibility and shadows of legitimacy in highly complex governance systems (cf. Alter and Meunier 2009; Bäckstrand et al. 2018; Zelli 2018), they examined whether such systems may make it too difficult for stakeholders to differentiate between the processes and performances of individual institutions. The climate and energy nexus, which brings together a large variety of institutions, provides such a complex environment.

The authors indeed found evidence that stakeholders in their sample did not adequately disentangle their legitimacy assessments of the individual institutions under scrutiny, since these overlapped in mandates and governance functions. Moreover, they found that a stakeholder's level of familiarity with an institution can be linked to a more positive assessment of legitimacy. Hence, there are reasons to believe that knowledge or valuing of an institution provides important markers or mental shortcuts for stakeholders when they navigate a very complex and densely populated subfield such as renewable energy.

These findings on legitimacy regard a small sample from only one of the three subfields of the climate-energy nexus. Future research has to show whether the institutional complexes on fossil fuel subsidy reform and carbon pricing exhibit similar levels of sociological legitimacy. On the one hand, one could expect such similarities, since, as we summarized earlier in this chapter, all fields exhibit comparable degrees of coherence and management. On the other hand, the governance systems on FFSR and carbon pricing are made up of considerably less institutions, which may qualify the impact of institutional complexity on stakeholder assessments.

Chapter 8 analyzed whether the institutional complexity of the three subfields affected various experts and stakeholders in how they assess the effectiveness of certain institutions. The analysis revealed many similar assessments across the subfields, in spite of the institutional differences among them, e.g. the much higher number of institutions for renewable energy than for FFSR and carbon pricing. The main disadvantages that experts and stakeholders in all three subfields highlighted were duplication of work, conflicting messages, and competition, which were all seen as obstacles in the way toward stronger synergies. On the other hand, institutional complexity offered more opportunities and venues to include a broader spectrum of actors and interests.

Potential explanations for these results may be related to our findings for legitimacy and call for further research. One could expect, for instance, that the social values, norms, and experiences that impact legitimacy assessments also play a role in effectiveness assessments. What speaks for this assumption is that the three dimensions of institutional effectiveness (output, outcome, impact) are also relevant dimensions for the output legitimacy of institutions.

An explanation for the differences found in assessments of effectiveness between the three subfields may be the overall level of complexity in a subfield as a more fundamental reason, and the respective explanations for coherence and management discussed in the previous section (Section 9.2.1). The position in a subfield in the climate-governance network or, for instance, the underlying constellations of interests, may not only shape the degree of complexity of that subfield, but also the (perceived) advantages and disadvantages arising from that complexity. All these potential connections point to the need for a groundbreaking research programme – one that provides scholars and stakeholders with novel, theory-driven analyses to explain and understand the complexity of a nexus and its consequences for legitimate and effective governance.

9.3 Recommendations

9.3.1 Enhancing Coherence and Management in the Climate-Energy Nexus

The ensuing recommendations are based on the empirical results of Chapters 4–6 of this volume. These chapters present three case studies on three major subfields of the climate-energy complex: renewable energy, fossil fuel subsidy reform (FFSR), and carbon pricing. The studies found that all three subfields, and the institutional complexes that govern them, are marked by medium coherence; minimal or low levels of management; and significant challenges, trade-offs, and conflicts. Given this similarity across the three subfields, our most important recommendations apply to all of them, and they are also of relevance for a wider universe of subfields within the climate-energy nexus that show similar levels of coherence.

First of all, improving coordination and building awareness of the activities of institutions is important for avoiding duplication of efforts and conflicting messages. Such conflicting messages may create confusion among relevant actors. They also provide alternatives for those actors that are opposed to certain policies and institutional efforts, helping such actors to 'shop around' for a different arena more suited to their preferences. Ultimately, conflicting messages may lead to conflicting impacts of institutions. This happens intentionally when institutions promote specific objectives at the expense of other goals. An illustrative example comes from the renewable energy subfield. There, the energy trilemma poses a challenge to all international institutions working on this area – namely to achieve three objectives simultaneously – energy security, energy access, and sustainable energy – that frequently conflict with each other. Many institutions settle for prioritizing one of these goals, and there is, in fact, currently a clear bias in global renewable governance toward clean energy access (Sanderink 2019). Against this backdrop, and to avoid conflicting impacts, there is a strong need for a more integrated, cross-institutional acknowledgement of, and approach to, the energy trilemma. In particular, the potential of renewables to address energy security must be recognized more widely.

Duplication of work is a serious issue given that the institutions have limited resources in terms of staff, budget, and expertise. Improving coordination and awareness of each other's activities is key – and it does not necessarily call for a single institution that acts as orchestrator or coordinator. Rather, strengthening and expanding existing inter-organizational coordination mechanisms and possibly establishing new ones could improve oversight and integration. To identify the most urgent governance gaps, it is crucial to keep track of which institutions and actors are performing which tasks in which part of the world. Such a clearinghouse approach or information hub could be modelled on, for instance, the Global Climate Action portal of the UNFCCC (NAZCA 2019). Based on such continuous and cross-cutting information, institutions could then ideally adapt or shift their activities or mandates accordingly.

Second, we need clearer definitions. All three subfields we analyzed suffer from differing and even conflicting interpretations of central concepts and norms, such as what constitutes renewable sources of energy, fossil fuel subsidies, and carbon pricing. These diverging interpretations may lead to inconsistencies and tensions when pursuing core objectives. For instance, the various and competing definitions of fossil fuel subsidies entail considerable differences between estimates of annual global fossil fuel subsidization – ranging from several hundreds of billions by the OECD and IEA (OECD 2018a, 2018b) to several trillions by the IMF (Coady et al. 2017). Such definitional ambiguity among international institutions and their members has allowed some countries to maintain that they have no subsidies at all, even though there are numerous methodological aspects on which these institutions agree.

A possible solution to this ambiguity could be some form of joint 'minimum' definition put forward by all intergovernmental institutions on FFSR. Such a minimal but flexible consensus could leave the door open to complementary approaches such as those of the IMF, which highlights the broader societal costs of government support for fossil fuels. Likewise, for the subfield of carbon pricing, diverging interpretations could be addressed by further specifying the norm of carbon pricing. A cross-institutional consensus could clarify that carbon pricing entails the payment of a non-trivial price for a significant share of emissions. Such a clarification could avoid conflicting signals and prevent carbon pricing schemes that only place a very low price on overall emissions. While such a clarification is strongly needed, it should still leave room for promoting both carbon taxes as well as emissions trading systems instead of just focusing on one of the two. Put differently, clarification is a balancing act. Too much norm or goal specification can come at the expense of a wider acceptance or support, particularly when competing definitions are rooted in actors' diverging preferences and worldviews.

Third, silo mentality should be avoided. Actors and institutions operating in the three subfields of renewable energy, FFSR, and carbon pricing should build stronger connections to other policy areas in the climate-energy nexus and beyond. For instance, the institutions in global renewable energy governance appear to compete with energy efficiency institutions over resources and visibility. Such competition can be in part prevented by separating opportunities for (financial) support and resources for renewable energy and energy efficiency, while simultaneously supporting collaborations between the two areas.

Another example of the need for stronger cross-area coordination is the importance of integrating fossil fuel subsidy reform within the UNFCCC (Skovgaard and van Asselt 2019). Moreover, in a broader sense of sharing experiences and spreading knowledge, institutions within the three subfields need to reach out to sectors and institutions beyond the climate-energy nexus. Many actors and institutions operating in other policy domains still need convincing of the urgency of a global energy transition toward renewables.

Fourth, certain institutions could act as orchestrators in order to facilitate these and other measures. To play such an orchestrator role, an institution would ideally have a broad membership and the convening power that comes with it, an extensive mandate, and a high degree of acceptance – qualities associated with major UN institutions – along with the organizational capacity to play such a role.

Within the renewable energy subfield, one candidate could be IRENA. Being positioned at the centre of the energy trilemma, it is one of the few renewable energy institutions that simultaneously promote energy security, energy access, and environmental sustainability concerns. Furthermore, IRENA's mandate to gather and disseminate comprehensive information would allow the organization to initiate a database on global renewable energy activities and to facilitate discussions on the benefits and shortcomings of various energy sources. Finally, having almost universal membership and being well-known as a focal point for renewable energy, IRENA can have convening power to invite institutions and actors from areas other than energy governance to join renewable energy discussions. All this said, a major limitation is that IRENA – like the International Energy Agency that could potentially also play such a role – is not a UN organization, which may affect its acceptance as an orchestrator.

Within the subfield of fossil fuel subsidy reform, the UN Environment Programme (UNEP) is well-placed to take over the function of an orchestrator. UNEP coordinates efforts to bring together multiple stakeholders to develop a reporting methodology for SDG indicator 12.c.1 on fossil fuel subsidies. It may also play a valuable role in improving coordination and coherence between the institutions working in this area (UNEP et al. 2019). Furthermore, UNEP benefits from having a universal membership. Nonetheless, efforts to avoid duplication of work could also be undertaken by (a subset of) institutions independently. Given the WTO's existing definition of a subsidy, parallel efforts to promote FFSR through this forum also help socialize a common definition of a fossil fuel subsidy (Verkuijl et al. 2019).

For the subfield of carbon pricing, the two most obvious orchestrators are the UNFCCC and the World Bank. Yet, both of them suffer from specific shortcomings. The World Bank is an institution in which industrialized countries have disproportionate influence compared to the UN institutions, which limits its legitimacy particularly in the eyes of developing countries. The UNFCCC has much greater legitimacy in this respect, but is constrained by the modest resources of the UNFCCC Secretariat and the often-protracted decision-making procedures.

Some of these concerns are also relevant regarding a more general orchestration role within the larger climate-energy nexus. Despite the UNFCCC's relatively large secretariat, it cannot be expected to coordinate the institutional complex on climate change. It neither has the mandate nor the organizational capacity to do so. The secretariat has, however, been engaging in light-touch coordination using orchestration as a mode of governance. It collaborates with other institutions and actors to provide platforms and data that could help to mitigate cross-institutional coordination gaps (Hickmann et al. 2019).

9.3.2 Enhancing Legitimacy and Effectiveness in the Climate-Energy Nexus

The feedback from selected experts in our surveys and interviews confirmed that greater cross-institutional coordination is also necessary to address some of the negative impacts of institutional complexity for the effectiveness in the climate-energy nexus. In line with the recommendations we outlined in the previous section, a more overarching and clear division of labour, reducing incongruence of messages, and facilitating greater information and data-sharing were all seen as key steps by interview respondents. They cautioned, however, that diverging interests will keep rendering such efforts difficult. Moreover, coordination attempts would need to be designed in such a way as not to overburden the institutions with bureaucracy and to instead strike a balance between greater harmonization and maintaining the autonomy of institutions. Otherwise, effectiveness could be hampered by ill-designed coordination attempts. Here, key member states that are members of several institutions may have a role to play to push for improved coordination.

Effectiveness ultimately implies that the output of the many governance institutions contributes to actual problem-solving – i.e. successfully scaling up renewables, providing stringent carbon pricing, and phasing out of fossil fuel subsidies. As Chapter 8 showed, it is difficult to isolate and trace back the impact of the three institutional complexes on such problem solving, either because that impact is limited in the first place, or because of limitations in data. Nonetheless, there is considerable scope for improving problem-solving in each of the three subfields we analyzed. This includes enhancing the effectiveness of the measures that have already been brought on their way. In the case of carbon pricing, for instance, this would imply a stronger concentration on the adoption of carbon pricing and on effective carbon price levels, e.g. in line with the recommendations of the High-Level Commission on Carbon Prices (2017) that suggests prices between 40 and 80 US $ per metric tonne of CO_2 equivalents.

Chapter 7 and its study of the legitimacy perceptions of five key institutions showed a significant variation across stakeholder groups. Against this backdrop, international institutions, first, need to apply different legitimation strategies to target different audiences and their varied interests. To be visible and reach various audiences in an increasingly crowded governance nexus, it is particularly important for international institutions to provide a credible picture of their purpose, procedures, and performance through multiple channels. The study differentiated the legitimacy assessments according to nine dimensions of institutional legitimacy, e.g. transparency, accountability, and procedural and distributive fairness. The responses on these dimensions provide a useful indication of which institutional qualities the five institutions need to communicate (and improve on) more extensively, in order to be perceived more positively by certain stakeholders.

Moreover, the study showed that stakeholders' familiarity with institutions matters for their sociological legitimacy. This points to the importance of outreach activities. Institutions operating in a dense institutional complex not only depend on performing their function well in order to be viewed positively, but also need to communicate and engage with a range of stakeholders, both members and non-members. Put differently, institutional complexity requires institutions to make additional efforts to improve perceptions of their legitimacy in two different turfs: first, in an intricate institutional landscape through cooperation and competition with other institutions that have overlapping mandates; and second, in a more complex communication landscape with outreach activities directed at different audiences.

9.4 Outlook

We conclude this chapter, and this book, with identifying urgent research gaps and future research avenues that we have repeatedly come across in our comparative and in-depth assessment of the climate-energy nexus. As Section 9.2 already includes our suggestions for further theory development on the analysis of a governance nexus, we concentrate our outlook on essential empirical questions of high relevance for researchers, policy practitioners, and stakeholders.

We structure our suggestions along our three case studies – renewable energy, fossil fuel subsidy reform, and carbon pricing – and the themes of legitimacy and effectiveness. There are cross-cutting aspects among them that point in the same direction, namely (1) to learn more about the causes of institutional complexity in the climate-energy nexus, and, more concretely; (2) to substantiate claims on the causal links between the degree of institutional complexity (in terms of coherence and management) and the implications of this complexity (in terms of legitimacy and effectiveness); (3) to identify conditions for successful management efforts and spill-over effects across institutions and levels and, while doing all this; (4) to go beyond renewable energy, fossil fuel subsidy reform, and carbon pricing, and expand studies toward other subfields in and beyond the climate-energy nexus.

For *renewable energy*, we need to know more about the interplay between renewable energy institutions and institutions from other energy-related areas, especially with regard to energy efficiency. Renewable energy and energy efficiency along with low-carbon technologies are key areas in decarbonizing energy systems and transitioning to fossil-free or carbon-neutral societies. However, as mentioned earlier, they more often than not compete in spite of their common vision. Moreover, and with a view to addressing the energy trilemma more appropriately, further research is required to explore how increased coordination efforts can take shape, and what role UN and non-UN organizations can play for this purpose, such as UNFCCC, UN Energy, IRENA, and IEA.

With respect to *fossil fuel subsidy reform*, one issue that could be examined in more depth is the interplay between limited-membership coalitions, such as the G20 or the Friends, and multilateral arenas. Such research could look into the extent to and conditions under which progress made in small-n settings can be integrated in multilateral decision making. Respective studies could draw on existing literatures in other areas, for instance on the role of regionalism in international trade governance (Baldwin 2014). A second avenue for further research is to parse out the influence of international institutions on the implementation of domestic FFSR policies. As Chapter 5 discussed, countries have taken on commitments to reform their fossil fuel subsidies through various international forums, including the G20, the UNFCCC, the High-Level Political Forum, and Agenda 2030. The extent to which these international commitments trickle down to domestic action needs empirical assessment, through examining which countries have implemented reforms following the introduction of such commitments.

Future research on *carbon pricing* could, when exploring causes and consequences of coherence and management, focus more on the role of individual members (see also Andonova and Mitchell 2010). Regarding research on causes, one could analyze whether or not overlapping membership between institutions facilitates inter-institutional coordination – in particular through specific officials or units of a joint member. For instance, it is likely that the same states were to a large degree represented vis-à-vis the UNFCCC by their environment ministries (see Skovgaard and Gallant 2015), vis-à-vis the World Bank institutions by their finance and development ministries, and vis-à-vis CORSIA by their transportation ministries. Exploring the impact of identical or different negotiators in different institutions on overall coherence could provide both empirical and theoretical contributions important to academics and policy makers alike. In terms of consequences, future research should focus on whether and how the presence of interaction mechanisms between two institutions influenced the decisions of actors to adopt carbon pricing. If the relationship between two institutions became more coherent, did this also make them more influential toward their members? Comparisons of such constellations across time as well as across different subfields could provide insights of great policy relevance.

Concerning the connection between institutional complexity and perceptions of *legitimacy*, this book has provided a novel empirical study on five institutions from the subfield of renewable energy. First of all, this type of analysis could be expanded to institutions from other subfields. Such a wider examination can provide comparative insights into how the different levels of coherence across institutional complexes affect perceptions of legitimacy. Second, further research is required to establish why different stakeholders assess the legitimacy of institutions differently in a complex institutional environment. A broader set of potential explanatory factors could be put to the test here to examine whether these differences ultimately stem from processes of socialization or whether they are rather due to functionalist or rationalist reasons. Third, while we asked stakeholders to assess institutions according to nine dimensions of legitimacy, the next step would be to evaluate the relative importance that each stakeholder type assigns to each dimension. This would help to capture the overall legitimacy of an institution in greater detail. Such a research avenue could also explain and understand how different dimensions of legitimacy relate to one another and which factors may lead to a legitimacy crisis. Fourth, and based on these insights, the role of specific legitimation and delegitimation strategies under institutional complexity merits further enquiry. Finally, our study focused on legitimacy at the micro level, i.e. for individual institutions. The next research frontier is to study how institutional complexity affects legitimacy for an entire institutional complex (meso level) or governance nexus (macro level).

With regard to *effectiveness* in the climate-energy nexus, our empirical findings indicate that output effectiveness, i.e. generating regulations and infrastructure, is high across the three institutional complexes for renewable energy, FFSR, and carbon pricing. However, behaviour-changing outcomes and problem-solving impacts of institutions were less significant in all three cases. Whether this is a result of institutional complexity, or perhaps a reflection of the limited authority of international institutions in these subfields, warrants further investigation.

Moreover, further research is needed to examine interactions across different subfields and how these affect the effectiveness of the climate-energy nexus (Sanderink and Nasiritousi 2019). On the domestic level, the three issues we scrutinized in this volume are often interlinked, e.g. in the sense that fossil fuel subsidies lower the de facto carbon price and make renewable energy less competitive. Likewise, renewable energy may provide a useful instrument for limiting potential negative consequences of FFSR and carbon pricing through energy access and rising energy prices. In turn, FFSR and carbon pricing may finance the expansion of renewable energy installations. Concerns about just transitions to a zero-emissions world (Newell and Mulvaney 2013) are relevant to all three issues and could benefit from addressing all three of them in a joint manner where possible.

Furthermore, studying other governance fields in a similar manner could provide a useful comparison to examine whether institutions in other policy domains have overcome some of the shortcomings of institutional complexity and why. Such research could help identify additional management options for international institutions to close governance gaps and enhance effectiveness in the climate-energy nexus. Finally, it should be noted that effectiveness as defined here does not capture the issue of efficiency. While the three subfields were all considered effective at the output level, future studies should also examine how economic efficiency can be improved under institutional complexity. This is important given strong competition over resources in different policy fields.

The governance of the climate and energy nexus, and the ability of its institutions to address key issues – such as renewable energy, fossil fuel subsidy reform, and carbon pricing – in a timely manner will have great consequences on people and the planet in the years to come. We therefore encourage scholars, stakeholders, and practitioners alike to address the theoretical, empirical, and societal dimensions of the governance of the climate-energy nexus sooner rather than later. We hope that the research presented in this volume provides a first and crucial step to advance this important research frontier and to ultimately help resolve one of the largest planetary challenges ahead of us.

9.5 References

Abbott, K. W., Green, J. F., and Keohane, R. O. 2016. Organizational Ecology and Institutional Change in Global Governance. *International Organization* 70(2), 247–277.

Alter, K. J. and Meunier, S. 2009. The Politics of International Regime Complexity. *Perspectives on Politics* 7(1), 13–24.

Andonova, L. and Mitchell, R. B. 2010. The Rescaling of Global Environmental Politics. *Annual Review of Environment and Resources* 35, 255–282.

Arts B. and Buizer, M. 2009. Forests, Discourses, Institutions. A Discursive-Institutional Analysis of Global Forest Governance. *Forest Policy and Economics* 11(5–6), 340–347.

Bäckstrand, K., Zelli, F., and Schleifer, P. 2018. Legitimacy and Accountability in Polycentric Climate. In *Governance Governing Climate Change: Polycentricity in Action?* edited by A. Jordan, D. Huitema, H. van Asselt, and J. Foster. Cambridge: Cambridge University Press.

Baldwin, R. 2014. *Multilateralising 21st Century Regionalism*. Background Paper, Global Forum on Trade, Reconciling Regionalism and Multilateralism in A Post-Bali World. Paris: OECD Publishing.

Bernstein, S. 2005. Legitimacy in Global Environmental Governance. *Journal of International Law and International Relations* 1(1–2) 139–166.

Biermann, F., Pattberg, P., van Asselt, H., and Zelli, F. 2009. The Fragmentation of Global Governance Architectures: A Framework for Analysis. *Global Environmental Politics* 9(4), 14–40.

Coady, D., Parry, I., Sears, L., and Shang, B. 2017. How Large Are Global Fossil Fuel Subsidies? *World Development* 91(C), 11–27.

Conca, K. 2006. *Governing Water: Contentious Transnational Politics and Global Institution Building* Cambridge, MA: MIT Press.

G20 (Group of 20). 2009. Leaders' Statement. Pittsburgh, United States. Available at www.g20.utoronto.ca/2009/2009communique0925.html.

Hickmann, T., Widerberg, O., Pattberg, P., and Lederer, M. 2019. The United Nations Framework Convention on Climate Change Secretariat as an orchestrator in global climate policymaking. *International Review of Administrative Sciences*. Online first. https://doi.org/10.1177/0020852319840425.

High-Level Commission on Carbon Prices. 2017. *Report of the High-Level Commission on Carbon Prices*. Washington, DC: World Bank.

Ikenberry, G. John. 2003. Is American Multilateralism in Decline? *Perspectives on Politics* 1(3), 533–550.

Karlsson-Vinkhuyzen, S. I. and Vihma, A. 2009. Comparing the Legitimacy and Effectiveness of Global Hard and Soft Law: An Analytical Framework. *Regulation and Governance* 3(4), 400–420.

Keohane, R. O. and Victor, D. G. 2011. The Regime Complex for Climate Change. *Perspectives on Politics* 9(1), 7–23.

Lenz, T. and Viola, L. A. 2017. Legitimacy and Institutional Change in International Organisations: A Cognitive Approach. *Review of International Studies* 43(5), 939–961.

Morse, J. C. and Keohane, R. O. 2014. Contested Multilateralism. *Review of International Organization* 9(1), 385–412.

NAZCA (Global Climate Action portal). 2019. Global Climate Action NAZCA. Available at https://climateaction.unfccc.int/.

Newell, P. and Mulvaney, D. 2013. The Political Economy of the Just Transition. *The Geographical Journal* 179(2), 132–140.

OECD (Organisation for Economic Cooperation and Development) 2018a. *OECD Companion to the Inventory of Support Measures for Fossil Fuels*. Paris: OECD.

OECD (Organisation for Economic Cooperation and Development) 2018b. OECD-IEA Analysis of Fossil Fuels and Other Support. Available at www.oecd.org/site/tadffss/.

Orsini, A., Morin, J.-F., and Young, O. R. 2013. Regime Complexes: A Buzz, a Boom, or a Boost for Global Governance? *Global Governance* 19(1), 27–39.

Raustiala, K. and Victor, D. G. 2004. The Regime Complex for Plant Genetic Resources. *International Organization* 58 (2), 277–309.

Rittberger, V. and Zürn, M. 1990. Towards Regulated Anarchy in East-West Relations. Causes and Consequences of East-West Regimes. In *International Regimes in East-West Politics*, edited by V. Rittberger. London: Pinter Publishers.

Sanderink, L. 2019. Governing Renewables Globally: Facing the Frames. Unpublished manuscript.

Sanderink, L. and Nasiritousi, N. 2019. How Institutional Interactions Can Strengthen Effectiveness: The Case of Multi-Stakeholder Partnerships for Renewable Energy. Unpublished Manuscript.

Schmidt, V. A. 2008. Discursive Institutionalism: The Explanatory Power of Ideas and Discourse. *Annual Review of Political Science* 11, 303–326.

Schmidt, V. A. 2017. Theorizing Ideas and Discourse in Political Science: Intersubjectivity, Neo-Institutionalisms, and the Power of Ideas. *Critical Review* 29(2) 248–263.

Scott, W. R. 2008. *Institutions and Organizations: Ideas and Interests*. London: Sage.

Skovgaard, J. and Gallant, J. 2015. *National Delegations to UNFCCC Conferences of the Parties: Who Participates?* Earth System Governance Working Paper No. 35. Lund and Amsterdam: Earth System Governance Project.

Skovgaard, J. and van Asselt, H. 2019. The Politics of Fossil Fuel Subsidies and Their Reform: Implications for Climate Change Mitigation. *Wiley Interdisciplinary Reviews: Climate Change* 10(4).

Stokke, O. S. 2001. *The Interplay of International Regimes: Putting Effectiveness Theory to Work*. FNI Report. Oslo: The Fridtjof Nansen Institute.

Tallberg J., Sommerer, T., Squatrito. T., and Lundgren, M. 2016. The Performance of International Organizations: An Output-Based Approach. *Journal of European Public Policy* 23(7), 1077–1096.

Underdal, A. 2002. One Question, Two Answers. In *Environmental Regime Effectiveness: Confronting Theory with Evidence*, edited by E. L. Miles, A. Underdal, S. Andresen, J. Wettestad, J. B. Skjærseth, J. B., and E. M. Carlin. Cambridge, MA: MIT Press.

UNEP (United Nations Environment Programme), OECD (Organisation for Economic Co-operation and Development), and IISD (International Institute for Sustainable Development). 2019. *Measuring Fossil Fuel Subsidies in the Context of the Sustainable Development Goals*. Nairobi: UNEP.

United Nations. 2015. *General Assembly Resolution 70/1. Transforming Our World: The 2030 Agenda for Sustainable Development.*

Van de Graaf, T. and Colgan, J. 2016. Global Energy Governance: A Review and Research Agenda. *Palgrave Communications* 2. http://dx.doi.org/10.1057/palcomms.2015.47.

Verkuijl, C., van Asselt, H., Moerenhout, T., Casier, L., and Wooders, P. 2019. Tackling Fossil Fuel Subsidies through International Trade Agreements: Taking Stock, Looking Forward. *Virginia Journal of International Law* 58(2), 309–368.

Wiener, A. 2004. Contested Compliance: Interventions on the Normative Structure of World Politics. *European Journal of International Relations* 10(2), 189–234.

Zelli, F. 2018. Effects of Legitimacy Crises in Complex Global Governance. In *Legitimacy in Global Governance: Sources, Processes, and Consequences*, edited by J. Tallberg, K. Bäckstrand, and J. A. Scholte. Oxford: Oxford University Press.

Zelli, F., Möller, I., and van Asselt, H. 2017. Institutional Complexity and Private Authority in Global Climate Governance. *Environmental Politics* 26(4), 669–693.

Zürn, M. 1993. Problematic Social Situations and International Institutions: On the Use of Game Theory in International Politics. In *International Relations and Pan-Europe. Theoretical Approaches and Empirical Findings*, edited by F. R. Pfetsch. Münster: Lit.

Index

Important note: all references to figures and tables are in italics

Institutions, Conferences and Regulations

People

Countries

Others

Numerical

For EU product safety concerns, contact us at Calle de José Abascal, 56–1°,
28003 Madrid, Spain or eugpsr@cambridge.org.

www.ingramcontent.com/pod-product-compliance
Lightning Source LLC
LaVergne TN
LVHW081934200926
848484LV00009B/2427
* 9 7 8 1 1 0 8 4 8 4 8 1 7 *